Getting
The Best
From Your Bees

David MacFawn / Chris Slade

Getting
The Best
From Your Bees

Outskirts Press, Inc.
Denver, Colorado

Getting The Best From Your Bees
All Rights Reserved.
Copyright © 2011 David MacFawn / Chris Slade
v3.0

Cover Photograph provided by Chris Slade.

Outskirts Press, Inc.
http://www.outskirtspress.com

ISBN: 978-1-4327-6646-7

Outskirts Press and the "OP" logo are trademarks belonging to Outskirts Press, Inc.

PRINTED IN THE UNITED STATES OF AMERICA

Contents

Figure 1 The Queen
Courtesy of Larry Wessinger, Columbia, SC USA

In writing this book Chris and I contributed equally and complementarily to the content and editing. We fed on each other's ideas and believe this led to a better book.   Dave M.

Beekeeping programs in the United States have historically focused on theoretical test questions and answers but have not included extensive practical hive manipulation techniques. This book explores practical in-hive techniques that should be learned by at least an advanced beginner level, beekeeping year 2, and fully utilized by the Master Beekeeping level. Other techniques exist, but we listed what we thought were the most important techniques. This is an advanced beginner level book. The authors are from the South

Eastern United States and United Kingdom. We hope that we bring this diversity and beekeeping commonality to this text. We are 'divided by a common language' and so have included a glossary of terms with translations if necessary at the end of the book.

The concept for this book was born at the Virginia Tech Winchester Virginia Research Center USA, for ways to improve management of the research colonies. One of the overall themes of this book is that you need to go with the natural tendency of the bees to be successful. You need to learn what "normal" is for a beehive.

This book was developed for use throughout the English speaking countries. Different local conditions may impact upon various techniques but these techniques are applicable in the UK and Southeast US and in many other parts of the World. The law is different in different states; for example the use of chemicals and antibiotics is banned in some countries but allowed in others. Some allow fixed comb hives and others do not. The laws are constantly changing and if we told you exactly what is allowed and what is not, the book would be out of date by the time it is printed. In the development of this book, the authors consulted on the techniques. If different ways to approach a topic were identified, both ways were included in the text and so noted.

When reading the manuscript keep in mind that English originated over 1,000 years ago with the English common people. That is why English is such a powerful language and is so widespread and accepted throughout the world. It does not have a central authority deciding what should be in or out of the language: the people decide that. Hence, you will find that our manuscript is a composite of the British English and the American English. We did that on purpose. Chris is more comfortable with the British English and I am more comfortable with the American English. We have both types of readers so we split the difference.

"All beekeeping is local" and what might work in Alaska may not apply in Australia. We advise beekeepers to join their local Beekeepers' Association and apply the techniques described in this book after discussion with more experienced members. The Association will also be able to advise on the current state of the law locally.

I was discussing bees with someone at work and mentioned what I learned years ago from Dr. John Ambrose the North Carolina State Apiarists, United States at North Carolina State University. John said to be successful with bees you need to understand enough about their nature such that you do things that are supportive of their nature and not against their nature.

The bees know better what they are doing than the beekeeper.
Watch and listen to the bees
David MacFawn

This book is dedicated to my Grandfather, Elgie Deberry, New Midway, Maryland, a Dairy Farmer and Beekeeper who understood the value of bees and started it all for me 48 years ago. David Elgie MacFawn July, 2008

Figure 2 David Elgie MacFawn
David Elgie MacFawn July, 2008

David has kept bees in Maryland (Dark German bees), Virginia (Italian), North Carolina (Italian), Colorado (Russian), and South Carolina (Italian and Russian Hybrid). David is a sideline beekeeper in the Columbia South Carolina, USA area. He is a North Carolina Master Craftsman Beekeeper, Co Founded the South Carolina

Master Beekeeping Program, South Carolina Beekeeper of the Year, Assisted Dr. Fell at Virginia Tech in the Virginia Master Beekeeping Program, Incorporated the South Carolina Beekeepers Association as a 501C3 Non Profit Corporation, Published several (over five) articles in the American Bee Journal. David is a beekeeper and has co-authored this book for the practicing beekeeper.

This book is dedicated to Arthur Worth who taught me beekeeping in 1977. He was keeping his head down in a shell hole in Flanders in 1917 when he saw bees working the poppies unconcerned by all the carnage. It was then that he decided, if he survived, he would become a beekeeper.

Chris Slade

Figure 3 Chris Slade

Beginning

Figure 4 Cliff Ward
Cliff Ward West Columbia, South Carolina, USA in his bee suit.

All my hives are made of wood since wood is natural and is said to 'breathe'. I usually paint the hives a light color that normally blends with the background. Whether the wood still breathes with a coat of paint on the outside and a coat of propolis on the inside is a question that has not been studied but I believe it does! However, in this picture the white hives contrast with the dark background. It doesn't matter that much unless you are trying to camouflage them. Cliff does not actually need his veil on yet since he is not going thru a hive.

The beekeeper should normally put on a veil and light the smoker before opening a colony. The veil helps ensure the beekeepers do not get stung in the eye which is likely to cause blindness. The smoker is to help control the bees. Sometimes a mist spray of water or 'liquid smoke' is sufficient, but it is wise to know the colony's mood swings before risking not using a smoker. Chris recently got a pair of very warm ears from opening a hive without smoke or veil.

Wear light colored clothing with a smooth texture. A smooth texture is more important than light clothing; however, both are important. Khaki and other light colors are popular. Some of the more extrovert use high visibility (to people, not bees) red lightweight all-in-one suits. They are easily washable and quick drying which is why those whose job it is to look for bee diseases tend to favor them.

Figure 5 Smoker
Note the volume of smoke and the metal smoker box. The metal smoker box is for carrying the smoker between bee yards lit and for placing the smoker in after the nozzle is closed after use.

Before opening a hive, ask yourself why you are doing so. Consult your notes from the last visit. Should you have brought something with you? Were there problems that need to be addressed? How was their temperament then? Is the queen marked? What will you expect to see if the colony is progressing normally? If there is a departure from your mental picture of a 'normal' hive you should investigate the cause. Beekeeper education is a good reason for opening a hive and will give 'added value'; idle curiosity or because you always open the hive on the third Wednesday of the month are not.

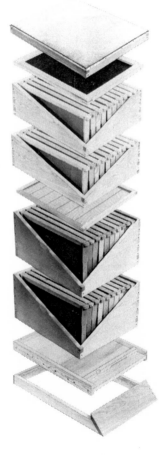

Roof / Telescoping Cover

Inner Board / Crown (Cover) board

Top super with shallow frames fitted with vertically wired foundation

First super ditto

Wire queen excluder, framed.

Upper brood box fitted with deep frames with side bars drilled for horizontally wiring foundation.

Lower brood box ditto

Solid floor with entrance block

Stand

Figure 6 Langstroth Hive
A Modern Langstroth Hive – Courtesy of Dadant (see glossary)

Figure 7 Eight Frame Hive
An Eight Frame Langstroth Hive – Courtesy of Brushy Mountain Bee Farm

Introduction

An experienced person can tell if a honey (nectar) flow is on by seeing bees darting back and forth with a sense of purpose. The more purpose the stronger the flow. Also, you will not see bees sitting idle on the front porch or bee beards on the front of the hive. This is another way of telling a honey flow is on, besides white new comb in a honey super and fresh nectar in the colony. If you see fresh white wax in the top super you should have put on another one a week ago! To tell this sense of purpose a person has to understand what a "normal" hive acts like.

Figure 8 Examining Hive Front
Cliff standing beside the hive front examining status of bees before lighting his smoker.

Standing to the side of the hive front, an experienced person can:

- tell if they are bringing in pollen indicating the presence of a laying queen or possibly laying workers.

- tell if there is chilled brood, (pupae in front of hive); chalk brood, (mummies in front of hive)

- tell if the hive is acting normally- normal means activity at the front, guard bees, no robbing (passive and active) other bees, wasps, yellow jackets, ants, etc.

- tell if the bee population activity seems low for time of day - know to look for poor queen or diseases or queen cells (you may have lost a swarm).

The job is easier if you have another hive in the vicinity with which to compare it. If one is doing something odd compared with the other, ask why.

How do you judge what is a 'lot of bees'? This picture may help.

Figure 9 Hive Front - a lot of bees
This hive has a lot of bees. Some of them are exposing their Nasenov glands and fanning – maybe there is a queen out on a mating flight and they want to make sure she finds her way home. Look for swarming signs when you go into the hive. This is one of Cliff Ward 's hives located at his extracting house.

Figure 10 Hive Front Minimal Bees
Looking at this colony, one would ask if it has swarmed, has diseases , possibly queenless or a drone layer; note also this hive has three supers on it. You need to look at the bee density in these supers.

When a hive is queenless it is noisy. You can confirm probable queenlessness if there are no brood/eggs in the hive. Also, the bees are restless. Dr. Fell (Virginia Tech, Virginia, USA) published a paper on this. The noise in response to queenlessness is primarily due to an increase in the number of bees scenting. Listen to the engine noise. Press an ear (preferably one of your own) to the side of the broodbox (outside that is) and tap on the box. If the hive is queenright you will hear a sharp buzz in response that dies away as quickly as it arose, but if the hive is queenless the buzzing will die away more slowly. Practice this before you need to so as to get your ear tuned in.

If the bees get very excited and you are stung heavily with alarm pheromone such that you can smell it or have stings in your clothing, you can wait until tomorrow to work the bees or start working hives at the other end of the apiary. You have failed to contain and stop the bees from getting excited with smoke (short puffs). These could be over defensive bees but even so, with the proper technique you should be able to stop defensive mechanisms. Before working another hive, you should smoke vigorously the clothing area "contaminated" with alarm pheromone. Alternatively, if you have a water source nearby, a quick rinse will get rid of most of the smell.

You should verify that something, such as small animals, is not causing the bees to be excited.

Figure 11 Two puffs from a well lit smoker .

Good beekeepers have the ability to keep their smokers alight to produce smoke when needed. This will assist in controlling the bees and allow you to use those short puffs of smoke to control the bees. You should light the smoker from the bottom by placing some combustible material in there, lighting it, and, after the fire is going well, placing more combustible material on top. Combustible materials include pine needles, wood chips, punk wood (partially decayed or rotted wood), etc. Old hessian potato sacks are good, particularly if they have been left out in the rain for a while and are starting to rot.

Once the end has been lit, it can easily be re-ignited on the next occasion with a mere spark from a cigarette lighter and a bit of puff.

If you are going from one apiary to another it is probably as important to be able to extinguish the smoker as it is to light it. Your insurance company won't look kindly on you if you set the car ablaze! Fresh leaves or a whittled plug to stop up the orifices are usually effective but may still leave your vehicle stinking of stale smoke. Try placing the smoker into a non-flammable container with a tightly sealed lid. It will go out on its own but will leave easily ignitable tinder in the firebox. If the journey is short between sites it might still be smouldering when you want to use it again.

While in Slovenia (home of the Carnica bee) Chris saw many hives opened but never saw a smoker , let alone a lit one. Usually no veils either. In one bee house the keeper ignited the charred end of a bit of stick and blew across it towards an open hive as you can see below.

Figure 12 Slovenia House Hive

When we opened a Langstroth in which the bees were thought to be a bit tetchy, a lady was asked to light one of her cigarettes and puff into the hive, which she did. Once was enough and after that she just stood around enjoying her ciggy. I think that's her in the dark glasses:

Figure 13 Slovenia: Franc Sivic looking for queen cells

If it is the swarming season keep an eye out for swarming preparations, i.e., queen cells, and have a plan in mind for what you will do if you find them. Never, ever, destroy a queen cell unless you are certain that the bees can replace it. We do not know what state, swarming or supersedure, the colony is in. If the queen cells are cut out or otherwise destroyed, it could mean the colony will be hopelessly queenless or take longer to make another queen cell if young enough larvae exist.

Bees know what they are doing (from the bees' perspective not the beekeeper's perspective), and after the new queen hatches, the beekeeper can assess if the new queen is satisfactory. Often, weeks in advance of queen cells appearing, the temperament of a colony will be a little sharper than you have come to expect when inspecting that particular hive as can be checked from your records.

Assess whether the density of bees is too great. If they are on top of each other they need more space.

Figure 14 Smoking the entrance

Cliff smoking the entrance. Note the handle cleats or pieces of wood over the hand holds. This increases the lifting surface area and makes it easier and more comfortable to lift a super of honey. However, it does make the equipment stack more difficult.

Smoke the entrance lightly if you are not looking for the queen. If looking for the queen, smoke the entrance even less before going into the hive. When smoking, you aren't trying to gas the bees, just to induce the guards to go and find some open stores to gorge. This takes time, so count slowly to 120 whilst putting your veil on and giving the hives subsequently to be examined a whiff (no more) of smoke.

Use slow methodical movements. Use short puffs of smoke as much as possible. Listen to the bees and watch their movements. The minute they start to increase their noise, smoke them a little. If a bee starts flying erratically, smoke them a little. Move hands around, not across a hive as the bees are attracted to rapid movements. If you move your hands over an open hive do so at half speed. When you get stung, it is usually as a result of a fault in your technique. There are those cases where you get stung by chance, for instance picking up a super and a bee is under the handhold or picking up a hive to move it and a bee is where your arm touches. Experienced beekeepers receive fewer stings than beginners because they usually have developed better techniques over the years.

Figure 15 Opening a hive
Cliff opening a hive.

Use a hive tool to break open the "seal" between supers and hive body. Go easy. Often frames from below will be stuck to the box above (if the bee space is violated), so ease up one side of the box with your hive tool, insert a wedge to take the weight and sort out the problem gently with the hive tool, maybe using a little smoke to contain the disturbance. If, occasionally, you apply some Vaseline or similar product to the parts of the hive where wood meets wood they will part much more easily without the jarring that disturbs the bees. You can make your own version of Vaseline by blending bees-wax and liquid paraffin (food grade mineral oil in the US).

Open the hive. There will be an inrush of daylight and a sudden change in the hive atmosphere (less so where mesh floors are in use). This will alert the bees that something is happening in which they should take an interest! Stand and watch for a few seconds, assessing the situation and the quantity of frames covered and the engine noise. Drift a little smoke across the top bars if thought necessary and maybe park the smoker on the upwind side of the hive, hanging it on the side of the box with the coat hook you screwed to the rear of the bellows, otherwise hook it onto your belt or pocket. You can use short puffs of smoke to direct the bees. Avoid using a bee brush as much as possible for hive manipula-tions. If you want to move bees off a frame to examine the brood it is better to shake them off, or cause them to move aside by gen-tly applying the back of the hand to a cluster. You can also blow on the bees and they will move. Don't shake a frame that has a queen cell or you may dislodge the potential savior of your colony from her food supply or damage her physically.

Figure 17 on the next page shows Cliff taking a second frame out. Note the amount of bees on top of the frames. This is a nice colony. It is a first year colony established in May in Columbia, SC USA and it is now mid August. The honey flow is first of April thru the first of June in this area. This colony is about 3 medium super frames short of having enough honey to get it thru the winter.

Figure 16 Marginal colony
A Langstroth hive before taking a frame out. This is a marginal colony that was started late in the honey flow (now is middle of May).

Figure 17 Nice colony: taking a frame out
Note also the ill-fitting bee-tunic (More obvious in Fig 24). If you can see up the sleeve, bees can crawl up it! Elastic or Velcro fastenings, properly adjusted, will avoid this problem.

Take out an end frame, look for the queen then lay it (the frame, not the queen) flat on the side of the hive furthest from you. When you are half way through examining the brood nest move it to cover the frames you have just replaced.. Generally you want to keep the queen in the hive on a frame as much as possible.

If a frame becomes jammed, possibly because of propolis, then leverage with the hive tool may well pull the frame apart. It is better to hammer the nails horizontally through the side bars into the top bars and not vertically as in the photograph above. David, normally nails the top frame as shown in this picture. It is the length of nail that holds the frame together in addition to glueing the frame. This doesn't apply to bottom bars as it is often helpful to remove them when renewing foundation (if used).

Examining frames, refer to figure 18. Note the brood showing in the lower left side of the frame and the bees covering the remaining frame.

Figure 18 Frame of bees
Generally you want to hold the frame upright and over the hive if the queen should fall.

Cliff is holding the frame "slanticular". This can cause problems with nectar dripping out (which might spark off robbing) or, where wired foundation isn't used, with the comb becoming loose in the frame. Try to develop the habit of keeping the frames vertical during examinations, this is especially true for Top Bar / Kenya Type hives. It means that you have to stand upright (which is much better for the back than continuously leaning over the hive). To see the other side of the comb: lower one hand until the top bar is vertical; rotate the frame 180 degrees; raise the hand again to restore the top bar to the horizontal position, but upside down. Reverse the process before putting the frame back in the hive. This is a long winded process to explain but it becomes second nature with practice. It is essential when using hives without frames or without foundation, to which we shall introduce you later in the book.

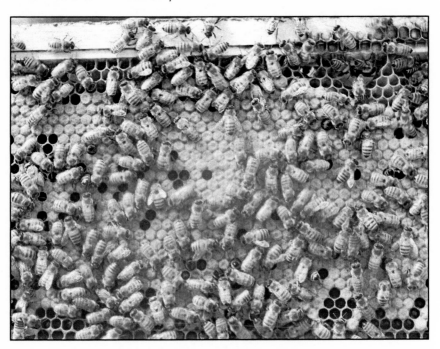

Figure 19 Reasonably good brood pattern

A frame of brood – This is a nice pattern with not too many holes.

Always seek an explanation for holes in areas of sealed brood. It might be benign such as a cell filled with pollen, neutral such as the wiring making the cell unsuitable, or a problem such as disease.

Examine the next frame. Make a mental note of what you see and compare it with the mental picture you established earlier of what you thought you would see. Place the frame back where it came from. Do not let it drop that last 1/2" or so into the gap. Remember you are burgling somebody's house, so don't set off the alarms! Continue looking at each frame all the way through the super or brood chamber to look for queen cells, proper hive layout, (brood in the center frames with a band of pollen , and honey in the frame corners, brood frames in the center with a frame of pollen then a frame of honey on the outside), disease, amount of brood, honey and pollen stores.

The queen should be marked on the thorax for easy spotting. The paint is usually applied while the queen is trapped under the mesh of a 'crown of thorns' cage. The dexterous can do it while holding the queen between fingers and thumb but try practicing on drones first!

International Queen Markings

Years	Color
1 & 6	white
2 & 7	yellow
3 & 8	red
4 & 9	green
5 & 0	blue

The mnemonic is: When You Requeen Gently Blob (her thorax)

Figure 20 International Queen Marking Colors

Observe relative calmness on the comb, i.e. bees stay on the comb or fly/migrate off the comb. You want the queen to stay on the frame and not fly out of the colony. If she does fly off; STAND STILL! She probably won't have flown for a long time, and possibly you have moved her since she became queen of that colony so she won't be familiar with the landmarks. One of the first things she will note as a reference point is a large white-clad object next to the hive – YOU! She may enjoy her flight for perhaps 10 minutes before returning home. If you are now standing next to another hive she may enter that one by mistake and be killed by the guards. Be patient. If there are two of you, watch each other's backs as I have seen a queen resting on somebody's back as she was standing in front of the hive.

You also want the bees to keep the brood covered, especially on a cool day. Observe how much capped brood there is and the pattern. This will tell you whether you have a productive queen and how many bees to expect in the future. It will also tell you, once you have got your eye in, if the colony has diseases. Notice holes in the brood. Excessive holes, or empty cells, will suggest that the queen may be short of sperm or mated with too few drones. Look to see eggs and larvae, the sign of a laying queen. See if the combs have honey in the corners and then a band of pollen. If not, in the fall/autumn you may need to assess whether to feed. Is the nest layout correct (brood in the middle with a frame of pollen and honey on the perimeters)? Have they swarmed? Does the brood, both sealed and unsealed, look healthy? You should have a clear picture in your mind of what healthy brood looks like. Any departure from that needs investigation and action.

Keep the hive open as short a time as possible. The amateur takes 20 minutes, the professional 2!

The rougher you are with your bees the more you will get stung.

Treat them gently and they will treat you gently in return. Avoid banging equipment. Place the supers back on the hive body; do not let them drop. Avoid mashing the bees around the edge of the super. Use a small wedge to ease the boxes back together while smoking the bees away from peril of being squashed. Or place the super on the brood chamber at a 45 degree angle then rotate to cover the hive. Don't jar the hives. Avoid this by lubricating with wax /oil (Vaseline in UK) where wood meets wood. Knocking the inner cover (crown board) to shake bees off into the hive is ok, just before closing up the hive. I do it and I still do not enrage the hive as long as I close the hive then. The rougher you are with your bees, the more the bees will "greet you" as you approach the beeyard or colony. Bees seem to remember rough treatment. Generally don't wear gloves unless you feel you really have to as they make you clumsy and insensitive. To gain confidence in going gloveless, cut the thumb and first 2 fingers off an old pair of gloves. A few drops of vinegar on your hands will disguise your 'enemy' scent.

Minimize drifting by arranging hives in a broken pattern facing different directions rather than in a straight line. One can also assist in reducing drifting by painting designs on the front of the colony in various colors. Use X and O pattern, and Blue, Blue-Green, and Yellow colors (bees cannot see red except as a shade of black) [9].

Honeybees do not care what the hive looks like as long as it keeps them warm, breaks the wind, and keeps the rain off of them and predators out. Keeping woodenware in good condition and painted is for the beekeeper. It keeps bees from coming out of unwanted holes in the hive whereby the beekeeper can get stung, (particularly if he uses his car for moving hives!) and lengthens the life of the wood/hive.

Keeping your Bees Alive and Thriving

There are several things the beekeeper can do to help his bees survive. Make sure the bees are dry, well fed, disease free, and good genetic stock. What is good genetic stock? Ideally the queens should have been bred and mated in the area in which they are to work. There should be a wide gene pool to ensure that there are numerous patrilines within the hive to cope with changing conditions and challenges. If you import bees from elsewhere your drones may be responsible for your neighbour's bees becoming bad tempered (and vice versa!).

Dry

Locate your bees in a dry location , away from valleys and other low points where mist and moisture may accumulate and frosts nip at night. Do remember, though, that they need access to water to dilute their honey or to cool the hive in hot weather. Place your colonies on hive stands to get them up off the ground. The colony should get morning sun to maximize their working day but, in less temperate regions, have shade in the afternoon to keep them at optimal temperature. Rising damp may rot the woodwork so place something impervious under the hive, like a piece of slate under corners to prevent this.

Well fed

Chris regards artificial feeding as extra unnecessary work that is unhelpful in the long term because it does nothing to select for bees that are in tune with and can thrive in their local natural environment. He hasn't routinely fed his bees for years and seldom loses a hive to starvation. He is also keen to ensure that the honey he eats, sells and gives away is the natural product of the hive gathered by the bees from flowers and is not partially composed of recycled cane or beet sugar.

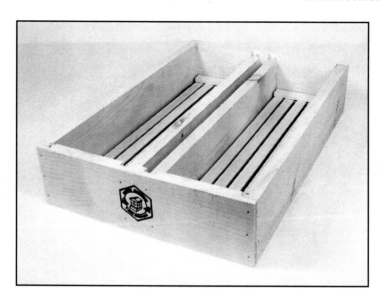

Figure 21 Wooden hive top feeder
Brushy Mountain Bee Farm hive top feeder. There is also a plastic version.

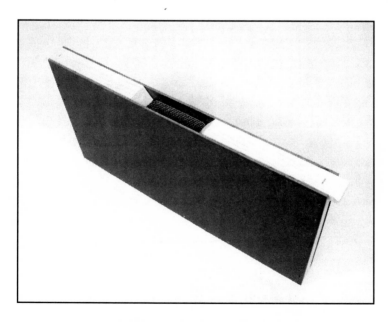

Figure 22 Frame or division board feeder
Brushy Mountain Bee Farm frame or division board feeder

Figure 23 Plastic frame or division board feeder
A Dadant frame or division board feeder

How good a fall flow is on average needs to be learned from experienced beekeepers in the area. If there is a weak flow in the
autumn, then you need either to leave honey on the hive from the
spring flow or feed. Alternatively consider whether the number of
hives competing for the available forage should be reduced. You
can do this by uniting your less good colonies (judged empirically
from your record cards) to the better ones. This can become a regu-
lar part of your bee breeding and improvement program. Leaving
honey from the spring flow is preferred since it is more natural and
contains minerals and nutrients that sugar syrup or corn syrup does
not. Also, if there is a dearth period between the spring and fall
honey flows, you need to leave extra honey on the hive. This is the
art of beekeeping.

A determination needs to be made in late summer and early fall
what your honey stores status is. Typically, with Italian types you
need 40-50 pounds in warm climates and 70- 80 pounds in colder

climates, but other bees, more suited to cooler climates, are often able to winter more efficiently.

Figure 24 examining a frame of bees

Note the honey in the upper corners of the frame. This brood pattern is "ok." but an explanation for all the holes in the sealed brood should be sought. It might be something innocent and beneficial like stored pollen , but, on the other hand there may be brood disease present.

If you are starting a new hive with a package or nucleus you need to assess shortly after the spring flow the colony status. In a Langstroth, if you only have 5-6 drawn and used combs, you need to start feeding or the new colony will not make it thru the winter. This is true only of colonies started late in the spring flow. Starting a new colony on sterilized drawn comb works much better and is quicker. You may even get a super of honey off the hive if the bees are installed early in the flow and it is a good long flow. This is in US conditions. In other parts of the World where the bee is a natural part of the environment, if they can't build up sufficiently during a normal summer to see them through the winter then they aren't worth keeping and the local gene pool would be better without them. However, keep in mind when the honey flow is in your area and when you establish your colony during that flow.

A fully established colony normally takes a year to establish. David tries to avoid feeding established colonies since he does not want to breed for feeding dependency for the colonies in his bee yard. Part of the art of beekeeping is leaving enough honey on the hive to overwinter and for dearth periods in the summer. The bee-keeper needs to avoid taking off too much honey that the bees will overwinter on.

When feeding, you can use a hive top feeder, a pail feeder in-serted upside down over the inner cover (crown board) hole, an entrance feeder inserted in the hive entrance (usually these are recommended only for water due to robbing), or a frame feeder where a frame is replaced and the feeder inserted. Feed sugar syrup (1:1 by weight or volume sugar and water in the spring and 2:1 by weight or volume sugar and water in the fall) or corn syrup, in that order of preference. You can also feed sugar on the inner cover (crown board). Slabs of bakers' fondant are also useful for winter feeding. They are either placed directly on the top bars with an eke to accommodate them or else put over the feed hole in the crown board.

Disease Free / Genetic Stock

Inspect your colonies once a month (during summer) for diseases. At the start of the season, look at pictures of the usual diseas-es to 'get your eye in'. Look for not only the traditional diseases such as American Foulbrood, European Foulbrood, Chalkbrood,

Sacbrood, Small Hive Beetles, but also Varroa and Tracheal mites. You'll need a dissection microscope to see Tracheal mites (otherwise known as Acarine) though, and a compound microscope to diagnose Nosema. Recently it has been recognized that there are two sorts of Nosema: Apis and Ceranae. They look so similar through a light microscope that the amateur can't accurately distinguish them. However, if you treat them (where the use of antibiotics is legal) with Fumidil and move them onto clean comb then it should not be too much of a problem. Nosema spores can be destroyed on comb and woodwork by exposure to the fumes of 80% Acetic acid. It is recommended that as a matter of routine all brood combs should be so sterilized when off the hive before re-use.

If you see bees with deformed wings, resulting from the Deformed Wing Virus transmitted by Varroa, this is an indication that Varroa is getting out of hand and should have been reduced before now. You will need to take effective action very soon or you will be in danger not only of losing the colony, but of spreading the mites further as the colony collapses.

'Colony Collapse Disorder' has been the object of much discussion. The symptoms (not everybody agrees what the symptoms are!) are that a recently prosperous colony is suddenly reduced to few or no bees; brood is present and, although the colony is undefended, no robbing is taking place. Chris has seen this situation on a couple of occasions in the UK in the mid 1990s and once in 2004. The hives had all done exceptionally well those seasons. The numbers of bees present were between 4 workers and a queen and maybe a couple of hundred and a queen. Despite it being fall/autumn and the time of year wasps are looking for sweets, as well as other hives of bees after a free meal, there was no robbing. The factor common to all was that Chris was aware of a severe Varroa infestation but delayed taking action in order to avoid contaminating the honey crop with chemicals.

Look at the brood pattern to discover whether you have a vigorous queen. You should not have excessive holes (cells where there is not any brood) in the brood pattern early in the season. The queen should also be laying a lot of eggs (up to 1500 per day) by early mid season. The queen will start with a lower rate of laying early in the season but by mid season/flow she should be laying 1500 eggs per day provided that the bees are able to bring in adequate food.

Use hygienic queens. A queen whose bees keep a very clean house or which pass the test of removing in a day brood killed by freezing a section of brood about the size of the base of a vegetable can (about 3" diameter) is said to be hygienic. Obtain dry ice or liquid nitrogen in the US or place the cells in a freezer for 24 hours then reinsert back into the hive. Hygienic queens produce bees that clean cells very well, quickly (<24 hours) and keep a clean hive.

When treating for Varroa, use the more recent "natural" medications such as oxalic acid and the thymol and formic acid based ones like Api-Life Var, Apiguard , Mite-Away II if they are licensed for where you keep your bees. Use IPM techniques like a screened bottom board (see glossary) and the shook swarm , which is particularly effective if timed right. Drone brood removal can be useful, but remember that you will want to get your queens well-mated so do this only on the colonies you don't mind eliminating from the local gene pool, or after you have completed your queen rearing for the season. Where you have a swarm, either natural or artificial, removal of the first brood to be sealed will take out a very high proportion of the mites present. This is because the mites, which prefer to spend most of their lives in snug cells with a plentiful food supply rather than riding piggy back on bees, will have been doing the latter for maybe a couple of weeks and so will take the first opportunity they can to enter a cell about to be sealed.

Figure 25 Screened bottom board

A Brushy Mountain Bee Farm screened bottom board. Notice the plastic insert on the bottom of the board. This is for taking Varroa samples. The entrance is unnecessarily deep and will allow the entry of pests such as mice. This entrance depth is standard in the US since this is the bottom of the nest and the bee space does not apply. In the US we use an entrance reducer for the winter. Some years ago Jeff Pettis - USA - studied the behavior of living mites that had fallen through the mesh. Where the floor was only half an inch below the mesh they all found their way back into the hive. Where there was a 2" gap none got back!

David tries to collect feral bees and incorporate their genes into his operation. Feral bees have evolved to survive without treatment that our domestic bees require. In my own operation, I have one colony that I have not treated for 3 years but it took a substantial colony loss to get that one colony. I plan to let the remaining colonies go natural. I will be looking for feral colonies to add to my stock. Be very careful here. Most commercial operations still treat since they depend on colonies for their livelihood. I can let my colonies go natural since I am a sideliner and do not depend on the bees for my sole source of income.

Experimental – Foundationless Frames

Chris has more or less given up using sheets of foundation and finds that:

1) the bees produce a variety of sizes of worker comb ranging usually from about 5.0mm in the middle of the brood nest to maybe 5.4 on the periphery (at Lat 50 N). Occasionally some 4.9 is found. Probably about 5.2 would be typical. The last time he measured some manufactured foundation it was 5.7. These figures relate to the width of a cell between parallel sides. The easy way to measure this is to put a ruler across a run of 10 cells and divide by 10. If you do a lot of such measurements it may be worth etching a Vernier type scale onto a sheet of Perspex or else cutting a truncated V shape from a piece of card to place over the cells to see which of the pre-marked measurements fits.

2) the slabs of brood seem much healthier with far fewer misses.

3) he is beginning to notice in some hives that the bees are not uniform in size, and

4) the hives with different sizes of bee in them seem to be doing better than those which have more uniform bees.

When a comb is simply old and dark and there is no suspicion of foulbrood disease Chris simply uses a hive tool or knife to remove it from the frame leaving maybe half an inch (more in the corners) to guide them. Otherwise he uses starter strips, which average about an inch deep, but are cut on the slant so one end is deeper than the other and these are placed alternately so a deep end is next to a shallow end. Often he uses the thin foundation supplied for cut comb to make the strips. It introduces the least amount of other bees' wax (and, we now learn, chemical contaminants) into the hive.

David uses a narrow thin piece of wood in the frame split top bar in place of foundation. Either way works well but both require the colony to be perfectly level. Note however, this frame comb is usually not strong enough tangentially to use in an extractor unless the frame was previously wired. Chris doesn't extract from brood combs but (when he can be bothered) uses nylon fishing line to wire the frames as the bees seem to tolerate it more readily than metal wire.

The one time to use a full sheet of foundation is when doing a Bailey change. One placed centrally in the new brood box which is placed over the old provides a bridge that queen and workers can use to get to the upper storey. The queen goes up and starts laying in the new comb even as it is being drawn, so keen she is on fresh comb to lay in. After a few days the beekeeper, having ascertained that she is there, can put the queen excluder between the 2 boxes, removing the lower one after 3 weeks, taking with it some sealed drone brood with its complement of varroa. Chris has found that a good time to make a Bailey change is when the first oilseed rape is appearing in the fields nearby, so the farmer, not he, feeds them.

Chris has normally had no trouble with combs straying off course except occasionally in his top bar hive, which has, of course, no frames. When persuaded back on course with a gently applied hive tool the comb resumes the straight and narrow.

An important factor in working hives without using foundation is that the hives must be LEVEL as the bees use gravity to guide them.

A spirit level is part of his kit as well as a few wooden wedges to make adjustments. On the odd occasion when he hasn't been able to find the level among the junk in the back of the car he has improvised with a bottle of drink.

Dave has gone to the Steve Taber type frame as another IPM measure. Steve updated Langstroth's idea of using round dowels.

Figure 26 Steve Taber foundationless frame

The bees draw the comb as pesticide-free as we can get (dependent on what chemicals are in the hive), and are of the optimum cell size based on what the bees want for survival, not man. There has been much discussion in the bee organizations as to what cell size foundation one should use. This solves that problem; the bees decide what is best for them. This style frame works for brood frames that are not spun in a honey extractor. Dave is now testing the frame for honey extracting.

We historically are using mainly worker foundation at the expense of the natural 14%-17% drones......we are losing diversity in our bees. Beekeepers wanted more honey and believed more work-ers would give them more honey. There is not much difference between the worker foundation amount of honey yield and using the naturally occurring mix of drone/workers. The reason for this is the bees control how many drones are produced. When nectar is plentiful, they produce more drones; when nectar is less, they produce fewer drones. When nectar is plentiful, they are able to also produce more workers. So the benefit of worker foundation is minimized. I believe this is but one of the problems we have introduced into our hives; this is one reason why Chris Slade and I have gone to foundationless frames.

Turning the swarming season to advantage

Swarming:

It is natural for bees to swarm. It is how they reproduce and survive. However, to produce a crop of honey, lots of bees are needed. If the colony swarms, approximately half the bees in the colony go with the swarm, thereby reducing your nectar gathering workforce, although there will normally be lots of brood left behind to make up numbers later. Hence, minimizing swarming is a goal of the beekeeper but swarming is a goal of the bees.

In order to avoid your bees swarming without having a constant battle with them that must be demoralizing and time wasting for both you and them, it is necessary for you to take some planned action. There are several plans open to you which, although presented as alternatives, might overlap to some extent.

a) Let nature take its course. If you catch the swarm you can use it; if not you might get paid to remove it from somebody else's property, and then use it. Most likely it will be lost. Generally they won't swarm again (although occasionally casts, or secondary swarms, are possible) and so, for that colony, your swarming problems are over for the year and you have a new queen that probably won't swarm next year unless you encourage her to do so by overcrowding. If you get to the colony before the new queen emerges you can take a harvest of excellent queen cells that you can use to re-queen other colonies to pre-empt them swarming. The down-side of this, though, is that you might select for early swarmers and that is your main criterion for selection.

b) Some strains are more prone to swarminess than others. One of the traits of Apis mellifera mellifera, the original Black Bee, for example, is that it is somewhat inclined towards supersedure late in the season rather than swarming. This is a useful trait which is worth watching and selecting for.

c) Do it to them before they do it for themselves. There are probably almost as many ways of making artificial swarms as there are beekeepers! Ok that's an exaggeration, but there are many variations on the theme and most of them work most of the time. The principle is to separate the queen and the bees that would have joined a swarm from those that would be left behind. You can then do a number of things with them depending on your goals; for example make increase, run as a 2 queen colony, reduce mite and disease problems, raise queen cells etc.

d) Requeen with a young nubile queen either directly or by introduction of a protected sealed queen cell. The latter can be done in a queenright colony.

How can you tell when a colony is getting ready to swarm ? Any time there are drones in the colony swarming is a possibility, so you can relax when there are none. That doesn't help much, so what other signs are there? Look at your records. You do record temper at each visit don't you? If there is a sudden unexplained deterioration in temper this might mean that the queen pheromone isn't as well distributed as they would prefer and if this persists they might want to replace her with a new one. This occurs often weeks before the first queen cells appear. If your records also show that the queen is two or more years old then swarming is a 'probable' rather than a 'possible' and you should do something about it. Often giving them more room and some wax -drawing work to do will postpone the event, possibly until later supersedure, or swarming next year.

Queen cups are another indicator, but a weak one as they are usually present in any vigorous colony and really are an insurance policy, giving the bees the option of using them if they want to. It's worth taking a peek inside queen cups to see what's inside, which is usually nothing but air. If you see an egg, then make a note to check again in a week, taking with you the wherewithal to take appropriate action. If you see a larva that is a couple of days old or more, then the bees can be presumed to have made their collective mind up to swarm or to supersede.

Which is it to be: swarming or supersedure? Again consult your notes. Did this queen come from a stock with a history of supersedure or from a swarm or swarm cell? If the cells are few (<4), they are on the face of the comb, there is a history of supersedure and a swarm wouldn't upset your neighbours, then it's worth leaving them, maybe giving them more space as an incentive to stay. If not all of these criteria apply then assume that they are intent on swarming.

Notice when you open your hive the number of bees at the front entrance and if you feel an increase in temperature or a wave of heat when you open the hive. This may mean congesting in the hive which, when combined with queen cells, may mean swarming .

Look down between the brood frames and notice whether the bees are visibly crowded. When you pull out the brood frames, are the frames totally covered and crowded? The goal is to maintain 5 bee/in^2 (Discussions with Dr. John Harbo, retired, Baton Rouge Bee Lab, Louisiana, United States.)

An experienced beekeeper notices when the drones first start flying. This is the approximate time that swarming will start to occur; starting with the time after the drones first start flying and mainly thru the honey flow period. There are several ways to reduce swarming; Demareeing[9] (keeping the queen in the brood chamber on a frame of brood and moving most of the brood frames up above the queen excluder) seems to be the only sure way. Some beekeepers cut out all the queen cells they see but invariably they miss one and the colony swarms anyway. Another method is to pull a frame or two of brood out of the colony and make up a nucleus or another colony from several colony frames.

When pulling frames out of overcrowded colonies, you need to pull enough frames from enough colonies to make new colonies of multiples of 5 – 10 frames. At least five frames seem to do better. I am not a fan of a four frame or less nuc, but am of a 5 frame or larger nuc. Basically in setting up the new nucleus or hive you want to "make it like a cell builder." i.e. stock it with not only brood and bees but also pollen and honey.

Colonies also seem to swarm in the spring / summer after a cool spell or a rainy spell. There's a reason for this. Bees usually die away from the hive, either from wear and tear, predators (swallows for instance) or simply old age. If they aren't flying they aren't dying, yet new bees are being born at that time of year at over 1,000 a day. Thus a few days non-flying weather (or no flowers worth the effort of visiting) means that the colony can very rapidly become overcrowded resulting in breakdown in distribution of queen substance, triggering swarm preparations.

They usually land on a nearby branch or tree before departing for their permanent location. Try putting some material (an old sock maybe) well impregnated with the scent of propolis and old brood comb wax on a low branch in the apiary where it would be easy to take a swarm. This isn't guaranteed to work but does often enough to make it worthwhile. Also place a 'bait hive' containing some old (disease free!) brood comb somewhere that you'll notice when bees are flying from it. They prefer a site a few feet off the ground rather on the low stands that are for the beekeeper's convenience. Often you will notice bees investigating the bait hive some days before a swarm arrives and this is your cue to do a check to see which hive is making swarming preparations.

As a beekeeper you want to monitor the colony and make splits or other arrangements prior to (instead of) the colony swarming. Remember, you want to keep the bees but make them believe that they have swarmed. Reducing brood nest congesting by pulling a frame or two seems to work well but not always. Demareeing seems to be the only fool proof way to prevent swarming.

Demaree Method of Swarm Control

The Demaree method is labor intensive which may not make it usable by commercial beekeepers. Basically the method instructs to destroy all the queen cells in the brood chamber, transfer all the brood frames above a queen excluder and/or a super of honey placing the brood frames in a brood chamber at the top of the hive. Place frames of drawn comb in the bottom brood chamber with the queen. Often one frame of brood will be left in the bottom brood chamber. After about eight days destroy any queen cells you find in the top brood chamber. The bees will emerge in the upper brood chamber and no bees will be lost.

Figure 27 Wooden bound queen excluder
Brushy Mountain Bee Farm queen excluder.

Figure 28 Double screen board
Brushy Mountain Bee Farm double screen board

Snelgrove Board

A Snelgrove Board is a board with a double screen and paired upper and lower entrances on at least three sides, but as can be seen above there are other versions. The beekeeper does similar operations as Demaree , i.e. move all the brood frames except one with the queen in a brood chamber above the Snelgrove board. Leave one brood frame with the queen in the bottom brood chamber. Fill the bottom brood chamber with frames of drawn comb/ foundation/ starter strips. The bees will believe they have swarmed and the brood will hatch out normally above the Snelgrove Board. Snelgrove's original system involves the bees in the upper chamber getting used to flying from one upper entrance but then being fooled by the beekeeper into returning into a lower entrance, thus reinforcing the queen and weakening the upper half of the colony. This is repeated at intervals using the various paired entrances in sequence. If timed correctly the colony won't swarm, but a new queen will be reared in and mated from the upper chamber. You can use her to requeen that hive or another one elsewhere. Check that she has a good brood pattern before doing so.

Figure 29 Snelgrove board
A Snelgrove board. I typically do not paint the inside parts of a hive to let the wood breathe.

Padgen

In the Padgen method, the hive that is intending to swarm has a frame of brood together with the queen removed and placed into another brood chamber filled with drawn comb or foundation. This new hive is put on the old site and the 'parent' hive containing the rest of the brood and bees moved to one side. The supers can go back to the old site too as the foragers will rejoin the queen and, as there is little brood to feed, can store what they bring in. There will be a shortage of young bees in with the queen and so they will give up their inclination to swarm. The old parent part of the colony raises themselves another queen and the field force comes into the new hive believing they have swarmed.

When the new queen has been mated and come into lay the hive can be transferred to the other side of the one with the old queen. The flying bees from the moved hive will then move to the original hive reinforcing their foraging force. You can repeat that trick some weeks later, preferably when you think a honey flow is about to start. If you used foundation or starter strips rather than drawn comb there will be lots of foraging bees with little brood to feed and nowhere to put all that nectar except the supers until comb is drawn in the brood box.

Cloake Board

A Cloake board (which was devised by Harry Cloake of New Zealand) consists of a rim that contains a slide for inserting a metal or plastic sheet to divide two parts of the hive, and provides a second entrance. It usually contains a queen excluder. The Cloake board is inserted with the metal insert out. The lower entrance is reversed and closed forcing the returning bees to use the upper Cloake board entrance. After 24 hours the metal sheet is inserted thereby isolating the top and bottom brood chambers. The queen should be in the bottom brood chamber resulting in the top chamber believing they are queenless. Grafted cells can be placed in the top or you could move frames of eggs from the bottom brood chamber to the top. You need to make sure there is, in addition to the queen cells, plenty of honey and pollen in the top.

Figure 30 Cloake board
Notice the queen excluder that is exposed when the sheet metal is pulled back.

The Taranov swarm.

A Taranov swarm is fun for the beekeeper and spectators and is very effective in separating potential 'swarm' bees from the 'stay at homes'.

This is a method devised by a Russian of that name back in 1947. He shook a swarm , marked the bees, reunited, and then, next day, shook the swarm again and found that the same (marked) bees made up the swarm and the remainder (unmarked) returned to the hive. From this it may be deduced that the bees that swarm are not a random selection but a 'deliberate' one.

The Taranov board is about as wide as the hive and maybe 4 feet long. It is placed sloping up to the hive entrance so the leading edge is level with the entrance but leaving a gap about as wide as your fist, say 4 inches. It is useful, but not essential, to provide a strip of old carpet or of roughly sawn wood on the underside of the leading edge for the bees to get a grip on. Most people just use whatever board of about the right size that they have around and prop it up on bricks, but a fancy, purpose made one with adjustable legs looks like this:

Figure 31 Taranov swarm

As may be deduced by the initials on the back of the borrowed tunic worn by the person in the foreground, this demonstration took place in the apiary of the Twickenham and Thames Valley BKA.

The beekeeper goes through the colony and (with one exception) shakes all the bees off all the combs onto the board like this:

Figure 32 Taranov swarm

Some will miss so it is good to have a sheet underneath the board. The exception is a comb with a queen cell on. This is gently checked for the absence of the queen and then replaced back where it came from. It is the practice of the TTVBKA to find and cage the queen and place the cage where the swarm is to form under the top end of the board. However, this isn't essential and I have never bothered. I rather wished I had on one occasion when I saw the queen take to the air but she soon returned and joined the swarm.

The bees soon troop towards the hive. The home-bodies jump the gap and return whence they came, but the queen and swarmers form a cluster under the board as can be seen in the following pictures:

Figure 33 Taranov swarm process

In the next one you can see where the queen cage is dangled.

Figure 34 Taranov swarm queen cage

Eventually, when they have sorted themselves out you will be able to take away the board complete with swarm like this:

Figure 35 Taranov Swarm

This can be hived in the apiary in the normal way, but place it at a distance from the parent and face it in a different direction or absent-minded foragers may return to their old home by mistake. The parent colony can be left to rear a new queen from the cell you left them or start a new one. Alternatively you could insert a protected cell from another colony whose genes you want to propagate.

Try this method for yourself – it's good fun!

IPM – Integrated Pest Management

What is IPM and why use it? When pests come along a pesticide soon follows. It works well and is affordable and everybody uses that and nothing else. Then, after a few years, it doesn't work any longer and the pest gets the upper hand, having built up resistance to the product used against it so continuously. In response, people dealing with pests are now getting crafty and are using a variety of methods in rotation; some chemical, some physical, some bio-technical. In this way it is far less likely that resistance will be built up.

The first part of an IPM program is to ensure the bees are well fed, dry, and disease free.

Well Fed

David subscribes to feeding bees the first year if the colonies were established late in the honey flow season, to establish the colonies but feed minimally in years thereafter. One of the arts to beekeeping is leaving enough honey on the hive for the bees to overwinter. The beekeeper needs to account for dearth periods, estimating winter conditions not forgetting how much honey it will take for the spring buildup.

On the other hand, feeding bees will tend to breed bees that are dependent on your feeding, as those which are incapable genetically of overwintering efficiently in your local area are not 'de-selected' by the winter. Chris doesn't feed his bees at all, as those nursed through their first winter will breed drones the following year and spread their unsatisfactory genes.

You will have to make up your own mind on this as your personal attitude and outlook will influence your decision making. Probably, if you just have a couple of hives at the bottom of your garden, you will regard them as pets and treat them accordingly; whereas if your stocks are into double figures you will want to achieve some stock improvement by selective breeding.

Dry

The hive should be sound enough to keep the bees dry but let moisture from consumed honey (18.6%) escape.

Screened bottom boards are excellent for venting out moisture. They also work for assisting in varroa removal and they should normally be used. When a screened bottom board is used there is no need for further ventilation and a reduced entrance can be used which is easier to defend from predators. An empty super can be placed between stand and floor in winter to reduce the draught.

Disease Free

Selecting the best queen with survival traits is the first priority. You want hygienic bees. Treating with the newer soft chemicals (like Thymol based) should be done only after the proper analysis is carried out like using the "sticky" boards for Varroa assessment. Varroa followed by American Foulbrood (AFB) and European Foulbrood (EFB) are three problems that should be focused on. Even for AFB, treatment should only occur after detection, then Terrymicin/Tylan and requeening with hygienic queens should be utilized. In moderate to severe cases, the bees should be shaken onto fresh foundation, the infected frames/comb burned and the equipment scorched with a torch.

In the UK both foulbroods are 'notifiable' diseases and the

beekeeper is obliged to notify the authorities when their presence is suspected. If confirmed, the Bee Inspector will take the necessary action, which, in the case of AFB will entail burning, then burying, the contents of the hive and scorching the inside woodwork. This has made AFB a rarity in the UK.

EFB may, in severe cases, be treated in the same way, but with light infestation the Bee Inspector, after discussion with the beekeeper, might choose to treat with antibiotics or by a shook swarm. In either event there will be a 'standstill order' preventing hives being moved from the apiary until the Inspector is satisfied that the disease has been cleared up. Burning used to be compulsory with EFB as with AFB and since that policy has been dropped EFB has increased markedly.

Varroa levels should be monitored and treated based on the "sticky" board local level recommendations. You should set-up a monitoring program with "sticky" boards inserted in the colony bottom boards or under the mesh floor at least Spring, Summer, and early fall. The early fall is very important and should be done just prior to the winter bees/eggs being laid. Dave believes it should actually be done the bee generation that will feed and take care of the winter bees (Dee Lusby's idea, USA). This is mid to late July in the Columbia, SC USA area. The winter bees are critical since Varroa will shorten their life resulting in the overwinter bees not living thru the spring and colony death.

In a 10 frame Langstroth, yearly pull the center two frames and put above the excluder. Push in the frames and put two new frames one on each outside location. When the center two frames are hatched, burn them. I do the comb rotation on a warm day in winter, when the bees are not raising young. The center frames do not usually contain much pollen or honey as these are more relegated to the outside frames.

For the most part, Dave believes in supplying the colony with a high quality queen and leaving it alone as much as possible and using the least amount of pesticide possible. Different medication should be alternated from one treatment to the next. Also, you will notice many treatments are for 21 or 42 or 54 days. These durations correlate to the bees' brood cycle. Maladies like chalk brood usually means you need to requeen and ensure you are not in a high moisture area. The queen seems to be the focus of a lot of issues.

There are basically two ways to swap comb/frames. Swap all 9 or 10 of them (11 or 12 in the UK National hive) and start the bees on fresh foundation (pesticide issues); or swap two frames out yearly usually the center two frames and place the two new frames, one each on the outside and removed after the brood hatches. However, if two new frames are introduced, when the wax is drawn, how badly contaminated will the new wax be? Would I have been better off replacing all of the frames and start from scratch? We do not know yet. The general thought in the USA is to just swap all the brood frames in a five year period; however in Europe the recommendation is for a much quicker brood comb replacement cycle – annually in Denmark for instance.

Currently, most of the bee literature is concerned about pesticide contamination with respect to killing or harming adult bees. Dave is more concerned about contaminated foundation wax harming eggs rather than adult bees. He believes this is the basis of some of our problems. The LD50 or lethal dose 50% level for eggs is probably at least one order of magnitude smaller and possibly two magnitudes smaller that currently set for adult bees.

Open mesh floors should replace solid ones as and when the opportunity arises as part of IPM. Not only do they allow a con-tinuous small attrition to mite numbers through dislodged living mites falling through the mesh, but they keep the hive much drier inside. This seems very much to reduce the incidence of chalk-brood. A sticky board or other monitoring tray can be inserted at intervals to make a rough assessment of mite levels in the colony by counting and comparing.

Don't just count the mites: look at them through a hand lens. You may notice that some of them have limbs missing, nibbled edges and even mandible-shaped dents in the carapace (which may be due to post mortem desiccation rather than mandibles). Keep a note of the numbers. Chris noticed that over time the proportion of damaged mites increased to about 40% suggesting that the bees were taking active measures to combat the mites. Compare colonies and consider breeding preferentially from those that damage mites the most.

An open mesh floor has other advantages too. When moving hives, there is no need for an additional ventilation board. The size of the entrance can be reduced to make it more easily de-fended against predators. A quarter inch depth of the entrance will keep most normal sized mice out (shrews may still get in though). Creatures trying to get into the hive for an easy meal can waste a lot of time and energy trying to get through the mesh.

'Hard' chemicals are still part of the tool kit available as part of IPM. The trouble is that they are persistent and contaminate wax for years. When you find it necessary and desirable to use hard chemicals, do so in the early fall after the honey crop has been taken. Mark that hive for complete comb replacement by the shook swarm method in early spring. Burn the old combs. It prob-ably isn't worth trying to save the wax for candles or polish, and, in any case, would you be happy sitting in a room with a candle burning wax containing some of the chemicals that have been put into hives?

A shook swarm is a very useful tool at the right time and is not as dramatic as it sounds. Simply, on a day when the bees are flying well, move the old brood box aside and put a new one in its place. It should have a full set of frames fitted with starter strips, or with foundation if you still use it. Remove the two centre frames. Go through the colony one frame at a time. Lower each bee-covered comb into the gap in the middle of the new hive and give it a brief but vigorous shake to drop off the bees. Repeat until all have been done and then replace the two frames in the middle. You will notice almost immediately that bees start fanning from the entrance to encourage stragglers into their new home.

You now have a broodless swarm with very few mites (most of them will be in cells in the old box) and so, when they have settled down a little, they could be treated with a 'soft', non-contaminating, method of your choice. What to do with the removed brood which is on combs that you consider to be contaminated? First, cut out and destroy any sealed brood in a way that won't allow bees access to it. Ideally the remainder should be dealt with in the same way; however, beekeepers are a parsimonious lot and are disinclined to 'waste' an asset. The combs with unsealed brood should be marked so that they can be recognized. PROVIDED YOU ARE ABSOLUTELY CERTAIN THAT THE COMBS DON'T CARRY ANY DISEASE WHATSOEVER, they could be added to a nucleus or other weak colony that needs building up, with a queen excluder between to prevent them being re-laid. They should be removed as soon as the brood has emerged, and the comb destroyed. Frames can be re-used after thorough cleaning. The safest course is to destroy all the brood when the swarm is shaken. Unless there is a honey flow at the time, the swarm will need a small feed to get some wax drawn. Don't overdo it: you don't want re-cycled bag-sugar ending up in the supers.

Some have expressed concern that the shook swarm method reduces slightly the subsequent honey crop; Chris questioned a Danish beekeeper at Apimondia who regularly uses this method and he says that in his experience this is not so. If you are concerned to squeeze every last ounce of honey from your bees then a compromise might be as follows: make shook swarms early in the year but instead of destroying the brood, add it to another colony set aside from the rest. This will build up very strongly but will probably have a massive load of Varroa. Besides using it for producing queen cells to requeen other stocks, you could wait until the main summer flow is starting and then make a shook swarm (this time destroying all the brood), transferring the bees to starter strips or foundation with lots of supers of drawn comb above the queen excluder. With no brood to feed and nowhere but the supers to place the nectar until new comb has been drawn you should get a good reward provided the flow materialises.

Supering

Top Supering : Top supering is where the new super is placed on top of the existing stack of supers that is on the colony. This is the way most beekeepers super. It requires less time, less disruption of the colony, and is easier. If concern over bee travel is an issue, an upper entrance under the top new super can be made. Both top and bottom produce approximately the same production results.

Figure 36 Top Supering
Top supering a colony.

Bottom Supering: Bottom Supering is where all the existing supers are taken off the colony down to the brood chamber, the new super is placed over the brood chamber, and the existing supers are placed on top of the new super. Theory says the bees have less distance to travel to fill the super. Bottom supering is useful when seeking a crop of Bee Bread as the bees will often extend the pollen arch above the queen excluder.

Figure 37 Supering Sequence
Taking the Top Cover off

Figure 38 Supering sequence: breaking loose the supers
Breaking loose the top super for bottom supering; for top supering just place
the new empty super on top of the stack.

Figure 39 Bottom Supering
Taking the top super off for bottom supering

Figure 40 Bottom supering using hive tool
Using a hive tool to wedge the two supers apart.

Figure 41 Brood Chamber
Dovetailed (called dovetailed in the US but it is really a box joint or a comb joint in the UK) 10 Frame brood chamber – Brushy Mountain Bee Farm

In the US all medications should be taken off the colony 30 days prior to the honey flow. This means in determining when to medicate, one should start with the average honey flow date, back up 30 days, then back up at least the amount of time the medication should be on the colony from the 30 day date. Also, honey supers can be placed on the colony any time after all the medication is removed (30 days prior to the honey flow). Better to have the honey supers sitting on the colony early rather than in your storage shed waiting for the flow. It should be noted that pulling all medications 30 days prior to the flow will not keep Fluvalinate / Apistan and Check Mite out of the comb and pollen.

In the English-speaking parts of Europe, beekeepers will need to have an explanation of the term 'honey flow'! Unless the apiary is next door to a field of oilseed rape for example, it is more typical for nectar to come in intermittently throughout the floriferous seasons depending mainly on the weather.

Colony Management

If, toward the end of the honey flow , the nucleus/package/swarm that you installed is only 4-5 frames out of 10 (Langstroth), and the nucleus/package/swarm was installed at the beginning of the honey flow, you need to assess the queen and the health of the colony. Sometimes you would requeen the colony. The beekeeper needs to determine whether the colony is requeening itself and, if so, do nothing.

If this were a colony in August or September, you would have to assess what to do to get it thru winter when, examining the hive, you find it is only approximately 5-6 frames. This is the "art" to beekeeping since you are guessing at what type of winter it will be and how much honey the colony will need to get thru winter. You are also assessing the queen with respect to whether the colony was established at the beginning or end of the spring honey flow.

However, you also need to assess what to do to get the colony through winter and whether this is desirable. Winter is a great method of using Darwinian means to eliminate from the local gene pool those strains that are unfit for the local environment. If you decide that they must be saved at all costs, combining with another colony; feeding; bringing honey stores from a healthy colony with surplus stores to the weak colony are among the ways this can be done. Assessing the queen early on is critical. After the nuc/package/swarm has been installed for 2-3 weeks, you need to assess the state of the brood; amount of capped brood, amount of holes in the capped brood, amount of eggs and larvae. Typically when the queen first starts laying she will have a significant number of holes (unused cells) in her brood pattern. However, after 3 weeks or so, these holes should diminish and you are looking for a level around 10-20% holes or less (this is based on practical experience, I know of no research that has determined this). You need honey and bees to make it thru the winter in addition to a laying queen and of course it must be disease free and with a low mite count.

In addition to honey, the colony needs winter bees to make it thru the winter. In the UK and Ireland a good colony should be able to overwinter on a single National brood chamber on its own stores.

Beekeeping Equipment

Figure 42 A beekeeper's equipment paradise

Of course, you don't actually need to use frames at all. These cork bark hives pictured in Galicia, Spain don't.

Figure 43 Cork bark hives pictures in Galicia, Spain (not legal in some states of the United States)

The bees come and go through small (easily defended!) holes in the cork.

Figure 44 Close up of cork bark hives pictured in Galicia, Spain

Not everybody has access to cork trees from which the rind is stripped periodically to make the familiar bottle stoppers as well as the cylinders pictured above. It doesn't take a lot of imagination to make a square version from planks, thus:

Figure 45 Galicia, Spain Hive
This is what is called a Bee Gum in Appalachia USA

Don't get the idea that all beekeeping in Portugal and Spain is primitive! The next picture shows a small part of the operation of a beekeeper who has about 6,000 hives and employs up to 40 people at times.

Figure 46 Extra equipment in honey house in Galicia, Spain

Record Keeping

Write up your record as soon as you have put the cover back on the hive. If you can confidently and unstung do so with veil off, resting the notebook on the roof of that hive give them full marks for temperament. Consider using coloured record cards, the colour indicating the year the queen was hatched and hence her age. Use the same colour scheme for marking the queens - the usual sequence is white, yellow, red, green, blue for the years ending 1,2,3,4, & 5 in that order and starting again at white for 6,7,8,9,0. Once in a lifetime you might find a queen that still has the original colour marking when it comes around again; the best I have seen was a queen that passed the fourth birthday before swarming. Longevity is a valuable quality in bees and is considered to be hereditary (as in humans) and therefore selectable. Unfortunately those who requeen by the calendar have no idea whether their bees have this quality. In other words, let your very best queens die of old age, possibly retired to a nucleus from which genetic material in the form of eggs or very young larvae can be harvested for a supply of daughter queens.

> For many years Chris has used a record card of his own design which is a simplified version of the BIBBA (Bee Improvement and Bee Breeders' Association) card. The cards are designed to be used in a 6 ring binder of the standard Filofax type and are 180 mm wide x 100 mm tall.

○ ○ ○ HIVE RECORD CARD ○ ○ ○							
QUEEN REF.		APIARY			MOTHER		DAUGHTERS
MARK							
DATE	Q.1	Q.2	Q.3	Q.4	Q.5	Temper	COMMENTS

Table 1 Hive Record Card

Chris gets a new set printed each year in the appropriate color for the year. This is not essential, but it makes the cards of old queens stand out. You are encouraged to copy the design or to adapt it for your own style of beekeeping.

Ensure that the printer prints the reverse side upside down and in register with the front side. You can get 2 cards to an A4 card and they need trimming to size with a paper trimmer. They work out at about 5 pence (10 cents) each. The holes are punched with a Filofax hole punch available from stationers, or else with an electric drill if you've mislaid the punch!

There are two additional cards that go at the front of the pack. The first is a reminder of the questions to be asked every time a hive is opened. They are to be found in Ted Hooper's "Guide to Bees & Honey" and are as follows:-

Q1	Has the colony sufficient room?
Q2	Is the queen present and laying the expected quantity of eggs?
Q3a	(Early in season) Is the colony building up as quickly as others?
Q3b	(Mid season) Are there any queen cells?
Q4	Are there any signs of disease or abnormality?

Q5	Are there sufficient stores to last until the next inspection?
Temper	scored on a scale 0 (they saw me off) - 5 no gloves, no smoke, no stings

Mostly the answers are noted with a simple tick in the box. This can be varied with a simple code, for instance an M under Q2 means the marked queen was seen or CB under Q4 means Chalk Brood.

○	○	○	RECORD GUIDE	○	○	○

Q.1 Has the colony sufficient room?
Q.2 Is the queen present & laying the expected quantity of eggs?
Q.3(a) (Early in season) Is the colony building up as quickly as others?
Q.3(b) (Mid season) Are there any queen cells?
Q.4 Are there any signs of disease or abnormality?
Q.5 Are there sufficient stores to last until the next inspection?
Temper on a scale 0 (bad) - 5 (good)

DEVELOPMENT OF BEES

Days from laying egg	Worker	Queen	Drone
Hatching of egg	3	3	3
Cell sealed	8-9	8	10
Emerges from cell	21	16	24
Mature & ready to mate	-	20	37

Table 2 Record Guide

There is room on this card for an aide memoir on Development of Bees in tabular form for both castes and genders and for the name and phone number of the local Foul Brood Officer (which is on the rear of the card).

The rear of this front card is shown in the diagram below.

Seasonal Bees Officer

....

....

....

....

Phone....................................... Mobile..

Regional Maff/ Defra Office

....

....

....

....

Phone....................................... Mobile..

Table 3 Record Information

Queen No.	Hive No. or location	Queen No.	Hive No. or location

QUEEN LOCATION

Table 4 Queen Locator

The second additional card is a simple list with the serial number of the queen and the location of the hive she is in. Note that it is the queen that is numbered and not the hive. This makes it easier to keep track of family trees on the female side.

The layout of a suitable queen/location listing is shown below and the card should have identical printing on its reverse for more colonies. If even more colonies are to be recorded then extra cards can be added.

There are various ways of measuring hygiene if one wishes. Simply, one kills a number of larvae within sealed cells by freezing or with a needle, replaces the comb in the hive and, 24 hours later, counts the number uncapped and removed and converts it into a percentage for scoring. Hygiene is thought by some to make the colony less susceptible to a number of pests and diseases.

Economy can be assessed by weighing the honey crop and deducting sugar fed. Alternatively, as over wintering economy is most important, one can weigh a hive when Michaelmas daisies are in flower and again when it is dandelion wine time and compare.

If you wish to improve your stock, the last line of the reverse side is the most important, particularly the last word - CULL.

HIVE RECORD CARD								
QUEEN REF.			APIARY				MOTHER	DAUGHTERS
MARK								
DATE	Q.1	Q.2	Q.3	Q.4	Q.5	Temper	COMMENTS	
ASSESSMENT:			TEMPER			1 - 5		
			HYGIENE			%		
			ECONOMY			1 - 5		
BEST USE OF THIS QUEEN:						BREEDER, PRODUCTION, CULL		

Table 5Hive Record Card Rear

The rear of the card allows for overall collation of all examinations to give a figure for comparison to other colonies. The card(s) itself (themselves) can be filed for permanent record purposes.

If you have the equipment you can download this card from Dave Cushman's highly recommended Beekeeping and Bee Breeding website. (http://www.dave-cushman.net/bee/newhome.html)

Queen Rearing and Bee Breeding

"Why bother with queen rearing?" you may ask. After all, the bees were managing this perfectly well for themselves for millions of years before humans did it for themselves! There is something in this, in that wild or 'feral' stocks have to fend for themselves and thus have strong natural selection pressure applied. They are the ones that are disease resistant, long lived, vigorous and are well adapted for the local seasons and forage. Fortunate is the bee-keeper that has a number of well-established feral colonies within drone flight of his mating apiary to obtain the genetic benefits accruing.

However, we keep bees in artificial circumstances, crowded much more closely together than they would prefer and often next to neighbours who might not appreciate swarms in their gardens. Also, swarming, the natural reproduction of the colony tends to reduce the honey crop from the parent.

So, being keen on controlling nature as much as he can, the bee-keeper feels as if he ought to do some queen rearing, particularly as all the books he has read tells him he ought. It also gives him the opportunity to improve his stock.

A review of some of the important literature is in order.

G.M. Doolittle, "Scientific Queen-Rearing," 2008 Reprint Wicwas Press, 1889, 1899 George W York, American Bee Journal, 118 Michigan Street. ISBN 978-1-878075-24-6

Doolittle is the original text on queen rearing. Doolittle lived in the later half of the 19th century when Langstroth discovered and invented the movable frame hive, the queen excluder came into being, and foundation was invented. Doolittle listened to and learned from a lot of beekeepers of his time. He put together the significant knowledge of his day into his work. Like a lot of us, Doolittle started queen rearing by using queen cells then developed his grafting technique to produce more queens in the time frame he wanted them. Doolittle understood the importance of good tested queens to the well-being, honey production and survivability of the colony. The difference between swarm and supersedure cells was understood by Doolittle.

The importance of a strong honey flow and plenty of pollen was critical for queen building. Initially, Doolittle removed the existing queen to force the colony to develop new queen(s). The importance of young nurse bees for queen rearing was understood. Doolittle understood moving the young cells to another location to let them develop, and he understood about switching colonies so a strong field force colony was available to produce a honey crop. "Shaving off the cells down to one-eighth of an inch for the septum of the cell to access young larvae (2-3 days after egg hatch, Doolittle said four days after the egg was deposited in the queen cell) for grafting was also understood. A division board feeder was invented by Doolittle. Also understood was, "the food fed to all larvae, up to the time they are 36 hours old, is exactly the same, whether the larvae are designated for drones, queens, or workers." Doolittle put 12 cups to a bar that a larva was transferred to prior to placing in the starter hive. After 10 days the colony was opened and the cells transferred to mating nuclei. Doolittle knew that bees should not be shaken off the frames but brushed to prevent damaging the larvae. The queen "nursery" or queen bank was developed by Doolittle. An original "fondant" or candy was developed.

Doolittle started queen rearing by using queen cells then developed his grafting technique to produce more queens in the time frame he wanted them.

Giles Fert, " Breeding queens," Translated into English by N. A. Riley and P. M. Greenhead O.P.I.D.A.ISBN 2 905851 11-2

Fert discussed in addition to the Doolittle method, the Miller and Alley Methods of raising queens. The Miller method is used by a lot of hobbyists and produces plenty of queens for a small operation. A fully drawn comb is cut with triangles on the comb bottom. This frame is inserted into a colony with plenty of honey, pollen, and nurse bees. The queen lays in the newly drawn cells in the triangles. This frame with less than 3 day old eggs is inserted into a queenless starter colony. For every three cells with eggs along the edge of the triangle, the eggs in the first two cells are destroyed by the beekeeper, leaving the third cell with an egg. This keeps the developed queen cell separate from the others so that it is easily removed to a nuc. Fert uses a starter and finisher type operation.

Harry H Laidlaw and Robert E. Page, "Queen Rearing and Bee Breeding," Wicwas Press, Cheshire, Connecticut, USA original 1907 ISBN 1-878075-08-X

Laidlaw takes Doolittle's discussion into the modern day. Laidlaw's book picks up where Doolittle and Fert stop. He discusses production of queen cells, mating, which Doolittle and Fert do not discuss as much, care of queens, controlled mating, genetics, selective breeding and genetic basis for queen resistance.

This is an excellent history of beekeeping. It has adetailed explanation of the queen development stages, Queen anatomy, Miller method, Alley Method, and Doolittle's grafting.

When is the best time to start queen rearing and bee breeding? Now! Look at your records. Are there any colonies that you would rather not be contributing to the local gene pool? Get them sorted out before you (or rather your bees) start queen rearing so that their drones won't mate with your new virgin queens. How you do this is up to you and your scale of operations and circumstances. A large-scale beekeeper might send them away on a paid holiday (pollination contract), but somebody operating on a garden scale might unite poor (but healthy!) colonies to better ones at the beginning of the honey flow or before winter having first removed the undesirable queens.

This will do nothing about your neighbours' hives (a neighbour in beekeeping terms is up to 10 miles away) so, unless you keep your bees in an isolated area you should join your local association and discuss cooperating with them.

Actual queen rearing starts as soon as you see drones flying. As outlined above, you need plenty of stores, especially pollen (protein for body building) and plenty of nurse bees. For a small scale beekeeper who wants to produce just one batch of new queens then probably the simplest way is to make an artificial swarm with the old queen on the same site with new foundation or starter strips, then come a queen excluder and several supers followed by another queen excluder incorporating a top entrance, the old brood box with the stores, brood and nurse bees. There will still be some contact with the colony beneath, ensuring access to fresh stores as necessary but sufficiently remote to prevent the queen's pheromones inhibiting queen rearing.

They will start to draw queen cells. If the queen is the one you want to breed from, then let them get on with it. If not, destroy ALL the queen cells and keep doing so until the brood is too old to produce any more. Then introduce eggs/larvae from your (or your BKA friend's) selected stock. The Miller method outlined above is as good as any. Grafting's ok too, if your eyesight's good enough, remembering that the larvae you graft shouldn't be much bigger than the eggs whence they came a few hours earlier.

Before the first new queen emerges (leave a time margin for error) split the top colony into a number of nuclei, each with a queen cell, if necessary adding frames with stores from other healthy colonies. Arrange the nucs in a circle around the site of the hive with their entrances facing inwards, removing the bottom box with the old queen to a new location, not necessarily very far away. The bees, including many from the bottom box, will distribute themselves fairly evenly between the nucs. The beekeeper goes along the next day and, by transferring combs (NOT the ones with queen cells!) between colonies makes the distribution of bees even more even.

Leave them there until they've mated and the new queens are laying, then make a formal assessment to see whether they match your requirements and are well mated. Cull those that aren't up to standard. For this reason you should rear a lot more queens than you need yourself. If you end up with to many for your own needs you can distribute them to friends in your BKA.

If you're already operating on a larger scale and feel the need to raise several batches of queens, maybe in cooperation with the local association, then choose your favourite stock (having objectively examined your records). Place an empty hive next to it. Go around some of your other hives with a travelling box and in it place from a number of HEALTHY colonies a frame or two of sealed brood with covering bees. Try to select combs with lots of pollen too. By the time you've moved them round a few miles in your car the bees will have united sufficiently not to fight! Transfer these bees and combs to the empty hive.

Wait for a few days and examine carefully for queen cells. Destroy them. During the few days, bees will have emerged from their cells, reinforcing the stock of bees of nursing age. Add a frame of eggs/larvae (W pattern if you like) from the hive next door and let them rear queens. As the queen rearing colony's bees hatch out and combs become empty swap them with the hive next door, transferring combs of sealed brood and of pollen as well as honey at first.

When you have harvested the first crop of queen cells, repeat the process as often as necessary, at the end of which you might like to unite the two adjacent colonies in time for a honey flow. It would be better each time to produce cells from a different queen so as not unnecessarily to restrict the local gene pool.

What do you do with the queen cells that are produced? Use them to re-queen your own colonies or distribute them to your friends in the BKA. The simplest way of re-queening is to place the cell in a cell protector (or wrap it in baking foil, leaving the tip exposed) and place it in the brood area of the colony you wish to re-queen.

Marketing

Bottling Honey

The jar that you bottle your honey in depends on what market you are trying to reach. The quart and pint jars sold by weight imply selling honey as a country item. The label describes any specialty honey. A hexagonal jar enables selling honey as an upmarket item. The slim queenline jar sets the standard. The slim design lightens the honey and makes it sparkle. Glass is usually preferred over plastic.

Most honey sold in the US is marketed in liquid or extracted form. Liquify crystallized honey by placing in hot non-boiling water for 30 minutes. This high temperature will destroy some enzymes that distinguish honey, a natural living product, from a collection of 'chemical' sugars. If you don't want to defraud your customers you should point out that they could buy HFCS which is just as good as this, a de-natured' honey, much cheaper! To avoid destroying the valuable enzymes, don't allow honey to be heated above 120 degrees F. Each time you reheat honey above 100 degrees, you lose some taste and you increase the levels of Hydroxymethylfurfuraldehyde, HMF. In the EU honey is not allowed to be sold with more than 40ppm of HMF. Frosting/foam on the side of a jar (harmless but unsightly) can be resolved by directing an office anglepoise lamp towards it.

As Peter Borst , United States, has written: "Anyone who has a taste for honey, knows that there is a difference between fresh unheated honey, moderately heated honey, and the industrialized syrup fobbed off on people AS honey. No amount of scientific analysis will ever diminish the value of fresh, unprocessed food versus the homogenized pasteurized treacle some people have learned to tolerate".

Clearly it is necessary to warm the honey to make it runny enough to strain and to bottle. This should be done very gently over low heat. The enzymes are destroyed at 120F and so for quality honey you must keep the temperature below that.

In the EU it is illegal to sell table honey with a HMF level of more than 40 parts per million, although there is no simple way for the ordinary beekeeper to determine accurately whether his honey is above or below that level. HMF is a natural breakdown product of sugars and increases over time and temperature, so honey that has been kept in store from a year of glut may not be marketable in a year of dearth unless it has been kept at low temperatures. There is little logic in this position as a pot of strawberry jam (made by boiling the fruit in sugar) may have a HMF level of more than 500 ppm and is perfectly wholesome. In the UK, moisture content can be up to 20% or 23% for Ling (Calluna vulgaris) honey.

Labels (US)

Labels describe the content of the honey.
1. Comb   2. Liquid   3. Cut comb   4. Chunk   5. Spun or Creamed

Other liquids might include Mercury for example, or molten Iron. Honey is denser than water. Marketing. 1-quart jar is not 32 oz, but 44 oz OR 1.247 kg.by weight.

A 1 pint jar is not 16 oz, but 22 oz or 623 grammes. US pints and quarts are smaller than the UK. (The imperial version is 20 imperial fluid ounces and is equivalent to 568 ml, while the U.S. version is 16 U.S. fluid ounces and is equivalent to 473 ml).

Honey labels must have name, address, phone, whatever makes you findable, and the weight. For jars less than 1 lb show oz and grams, For jars less than 4 lb show pounds & oz and kg, For jars greater than 4 lb show pounds and kg.

Moisture content must be less than 18.6 %. This is the standard set above which honey may ferment.

You may use other marketing labels: HONEY should be largest word, Pure Honey, RAW (not heated above 100 degrees F), Blue Ribbon. The most common customer is female 25-49 years old; the more educated the greater the honey consumption. Older people tend to prefer stronger flavours than young ones.

For business greater than $50,000, it is mandatory to have nutrition label (shows no fat).

Commercial honey is put through filters and removes pollen , wax, & some flavor. In the EU filtered honey is labelled as such.

Honey in food / cordial production

You cannot replace sugar with same amount of honey. For each cup of honey used, reduce the liquid in the recipe by ¼ cup and add about ½ teaspoon baking soda, and reduce oven temperature by 25 degrees to prevent over browning...

Honey is denser than many other liquids; a 12 ounce jar equals one standard measuring cup.

Distribution channels

When you first start, Farmers' and Flea markets are an excellent outlet to sell your honey. You may consider setting up shop on a certain day and rent the same booth repeatedly. This way your customers will know how to find you and you can set aside a certain day to sell your honey and other bee products. Many churches use pure beeswax candles in their services. In Slovenia, proud home of the Carniolan bee, every market place has several family market stalls selling hive produce of every sort, even including propolis tinctures.

Figure 47 Slovenian Bee Products Stall

As you grow you will develop other channels such as small stores and supermarkets or health food stores. Get to know the owner or manager of the store and ask them what you can do to ease and improve the selling experience. It may be that the manager wants a certain size jar or squeeze bear. Or he wants a certain style label or help stocking the shelves. Set up a system where you are notified when the honey is getting low, or better yet, set up a time when you will come back to ensure your honey is being presented properly on the shelf, i.e. no jars are leaking or granulated.

It takes time to support your channel/outlet customers. Ask your customer if he wants you to retail price the honey and purchase the necessary pricing gun to do it. Also, find out if the customer wants a bar code on the honey for the price. You will have to decide whether the revenue received from a certain volume of sales will support the extra barcode cost.

Skill sets ....The Marketing /distribution skill set is different than a beekeeper/producer. As a producer, you need to decide if you will market your honey yourself or sell it wholesale and let someone else market your honey. If you market your own honey there will be distribution and marketing expenses and it will take time away from your beekeeping. This time typically is during the off bee-keeping season and is mainly in the September to December time frame in the USA. However, you will pocket the marketing profit also and make more revenue.

Production and Marketing cost ....

You need to know your cost to decide whether you are making a profit or not. Production records should be kept as should Marketing cost records. You are driving blind if you set your price market based and do not know your costs. You could be wrapping your own dollars around every jar sold if you are not careful.

The supermarket can be used as a base price but local honey should command a higher price. You need to know your production & marketing cost but set your price market based, then calculate your profit. Market based means what honey is currently selling in the market place whereas cost based means setting your selling price based on your cost plus a percentage mark up. If you do not set your price market based you could be leaving dollars on the table.

A lot of states have a promotional program to sell "local" state produced agricultural foods. These are excellent programs and can greatly increase sales.

Stocking shelves

The honey on the shelves needs to be examined periodically for breakage, granulation, and other appearance issues. Most customers 'buy with their eyes' and so it is worth while withdrawing from sale jars of honey that don't look their best and also to advise the shop keeper on stock rotation and of the optimum conditions for keeping honey in merchantable condition. It is better to supply him with small quantities often than to sell a lot at a time and have it granulating on the shelves. As stated, people buy with their eyes, so remove and replace any that are not good enough for the show bench. Don't sell the shop keeper more than he can shift/sell in a month. A lot of retailers want you to price your honey with a lot of supermarkets wanting a barcode on the jar. The barcode is added expense but may be necessary to sell in that outlet.

Figure 48 Shelf Display of Honey

Notice the display and that the honey is located next to other sweeteners. Also notice how some of the honey containers enhance sparkling honey, whereas others do not. It is a big selling point to have honey that is clean, without foam. Most North American markets prefer light honey whereas European and other markets prefer dark honey. Also, note how the shelves are nicely fronted or faced, i.e. jars to the edge of the shelf and the shelf full.

Figure 49 Sparkling Honey Shelf Display

Notice how this honey sparkles and it is in squeeze bears and also jars. Also notice the honey dipper hang tag. You can make additional revenue selling honey related products that consumers can use for their honey.

Figure 50 Sue Bee Honey

SueBee is one of the largest players in the US honey sales. Notice how the information on the label is laid out.

Figure 51 Southern Home Honey

Notice how the designs on the labels support the honey and that the animal depicted on the label has almost no resemblance to a honeybee but is a cartoon of a smiling child in fancy dress, thus designed to appeal to shopping mothers!

Figure 52 Golding Farms Honey

Notice how the honey is being sold as All Natural Fat-Free Food and that the irrelevant runners on the label hint that eating the product will enable you to do the same.

Figure 53 Nutrition Label

The nutrition label was added directly on the front main label. Notice the Bar Code.

Figure 54 Organic Honey

The types of jars are interesting here. Also notice how organic is coming into the market place. No information is given as to how the producer knows the honey is organic in a country where genetically modified crops are so widespread.

Figure 55 Cliff Ward Local Honey

This is local honey, actually some of Cliff Ward's honey. Cliff placed reliquifying instructions on the label. Sometimes the label will come off when placed in warm water. Also notice the Certified SC Produced Sticker on the jar's lid. For many people wanting honey with local pollen for allergies, this is a big seller.

Figure 56 Local Honey

Cliff sells pints and quarts in a Health Food Store.

Products of the Hive

Just a few tips will be discussed for the hive products. These are discussed at length in other bee books.

When determining what to produce, the beekeepers need to look at not only the costs but also the revenue for that particular product. Only by looking at both the cost and revenue can you get a feel of what to produce. These costs should include production cost, waste, travel cost, equipment cost / depreciation, labor (a lot of beekeepers forget about labor but if it takes twice as long to produce one type over another type product, your time is worth something), overhead (building, utilities, land, etc.). On the Revenue side one needs to look at what price the product can be sold for and also marketing cost such as sales, distribution, specialty jars, labels, etc. also "defective merchandise", (that honey that crystallizes on the shelf)

Figure 57 Bee-O-Pac Honey
Dadant Bee-O-PAC comb honey

Figure 58 Bee-O-Pac Ready to Sell
Courtesy of Dadant

Figure 59 Bee-O-Pac Super
Courtesy of Dadant

Figure 60 Hogg Half Comb
Courtesy of Dadant

Figure 61 Ross Round
Brushy Mountain Bee Farm Ross Round Comb Honey Super

For instance, in the case of comb honey such as Hogg Half Comb vs. Ross Rounds vs. Bee O Pac. Often if the flow is not strong or ends, you will have sections that are not filled or partially filled. That means you have Hogg sections that you cannot reuse (they are dirty and even the clean ones mean there is a double layer of midrib wax), Ross Round sections, and BeeOPac sections. You need to determine the cost of each (Hogg is more expensive than Ross Round) on the cost side. However, on the Revenue side, BeeOPac brings in the highest price per pound since it is smaller sections.

Chris simply allows his bees to build their own comb with just a tiny starter strip of wax. The best is then cut to size and sold in plastic cartons with clear lids to show the customer how beautiful it is. Offcuts can go in jars for 'chunk honey' and the remainder put in the press. This is the type of analysis that needs to be done for each product. You need to know your business and what you are getting into. You should make your mistakes on paper in trial runs not in real life. It costs way too much money.

The overhead cost should include things like product insurance and beekeeping liability insurance. I inquired about the cost of product and liability insurance in the US a few years back. It was $800 - $900 for a sideliner type operation which means I had to produce a lot of product for my profit margin to cover the cost of the insurance. i.e. to break even $0 = Total Revenue − Total cost including insurance. In England, the cost of such insurance is included in the annual subscription to the BBKA. In the current year, Chris' total subscription for the local, the County and British BKA, including Bee Diseases Insurance was £31 - say $60.

a) HONEY (extracted, comb, sections, pressed, creamed, soft set etc) You can either sell it in bulk as soon as extracted to a honey packer or cooperative, which means you won't get much money but will have more spare time, or you can process and sell it yourself.

To produce a particular type of honey, you have to move your colonies to a location with that particular flora and at the correct time of year, when the plant is blooming. You may consider putting the colonies on a trailer and leaving them on the trailer if you plan to move the colonies again. This will save unloading the truck then reloading the colonies back on a truck after the honey flow is over.

Figure 62 Extracted Honey

Bees can be removed from extracting supers by brush, fume boards with chemicals, and bee escape boards. Brushing works very well but takes a lot of time. Brushing also only takes one trip to the bee yard. Fume boards with chemicals such as Bee Go, or Jim Fischer's Bee Quick take only one trip to the beeyard and are relatively fast but you have the chemicals to deal with. Bee escape boards are the more environmentally sound method in that they do not require chemicals. It takes a day or two to remove with the boards but they work very well.

Figure 63Frame of Honey
A frame of honey.

Figure 64 Frame of Honey
A nice capped frame of honey. This frame would take some work to remove
the caps since the caps are recessed below the top bar surface.

Figure 65 Frame of Honey
A nice, fully drawn, frame of honey with a few bees. Note that Cliff's tunic
is slack at the waist and wrists which would enable bees to get at him if they
were in an attacking mood. A less experienced beekeeper, or one with less
docile bees would be wise to ensure their tunic was bee-proof!

Figure 66 Shallow Frame of Honey

A nice shallow frame of honey. This was taken inside a honey house away from the bees. This honey house or extracting house is not air conditioned (cooled) so that the keep the honey remains thin and extractable.. The cappings on this comb are wet with honey, which means that it wouldn't win a prize at a honey show. The comb also appears to be less wide than the side and top bars, making it less easy to uncap. This suggests that the frames are too closely spaced.

Most motorized extractor manufacturers recommend that the extractor is bolted down to the floor. For large extractors this works very well. For radial extractors in the 10 to 12 frame range you can put the extractor on wheels to move around and when extracting put the wheels in caster cups. For a hobbyist or sideliner this also works very well. Chris puts his 10 frame radial hand operated extractor on a board with castors and finds that the permitted motion allows the unequal forces from the rotating frames to resolve themselves without much movement of the extractor.

Figure 67 Bottling Types of Honey
Chunk honey put up in jars. The beekeeper can charge a premium for this.
Many people believe this is the "real honey" and will pay more.

A sideline or larger operation should consider double buffering the extracting process. This is where two smaller extractors are used. One extractor is extracting the honey while the other extractor is being loaded.

Figure 68 Uncapping Tank
A wax uncapping tank with cappings in tank. The wax cappings from a frame of honey are cut off into this tank. The frame is rested on the horizontal bar while the uncapping knife is applied. The frame is placed in the tank prior to placing in the extractor. Any honey that drips from it is directed via a strainer mesh into the bucket below.

Figure 69 Heated Knife

A heated uncapping knife in the foreground and extractors in the background. A heating element with a thermostat is in the knife's blade. This makes it easier to uncap a frame of honey, although a simple carving knife with a serrated blade will work very well. The heated knife does melt the wax which means it will damage the honey touching the knife.

b) WAX is a valuable product and you should aim for £1/ $2 an ounce minimum for clean wax, much more if you have added value by making polish, candles, ornaments, cosmetics etc.

c) POLLEN A very youthful looking pensioner I know buys it by the kilo! It is becoming increasingly popular as a health food.

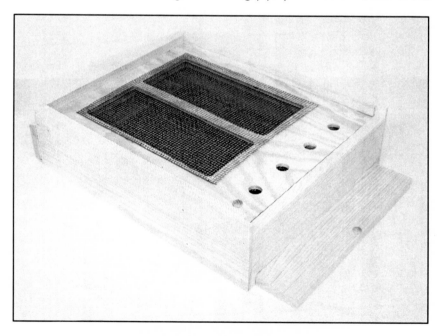

Figure 70 Pollen Trap
Brushy Mountain Bee Farm Pollen Trap

Figure 71 Front Pollen Trap
A Front Pollen Trap – Brushy Mountain Bee Farm

d) PROPOLIS At a recent Apimondia there seemed to be more stands and traders concerned with propolis than with honey! Its well-known therapeutic properties are making it increasingly popular (and therefore valuable)

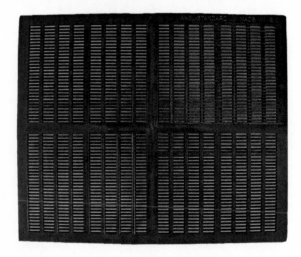

Figure 72 Propolis Screen
Dadant Propolis Screen

E) BEE BREAD This has to be the scrummiest product of the hive! A friend promised to send me a birthday present of some coated in dark chocolate. It never arrived so I guess she ate it all herself!

F) VENOM Recently I had a touch of tennis elbow (lateral epicondylitis) which I recognized from previous bouts that have lasted months and entailed wearing an arm brace and being injected with cortisone by the doctor. This time I applied the sharp end of a bee to the painful spot and left it there for at least 2 minutes to ensure the full dose. The sting seemed more painful than usual and the site went red and swollen for a couple of days. When the swelling was gone, so was the tennis elbow. Unless you are medically qualified or have a very good insurance policy it is unwise to diagnose other people's ailments or to treat them with bee-venom. There is a small market for collected venom but it is very specialized.

g) PACKAGES The UK generally does not use packages since they are quick and easy to spread diseases and acquire bees and their genes that are unsuited to your area.

h) NUCLEI. The people who make the most money out of beekeeping are those who sell things to beekeepers! So why not get a production line of nucs going? It can mesh very well with queen rearing and swarm control operations.

I) ROYAL JELLY Consider how little royal jelly there is in a queen cell, and yet the Chinese export hundreds of tons of it each year! Let them get on with it.

J) QUEENS The grass is always greener in somebody else's field. This is why beekeepers, instead of producing their own queens, suited to their own location, spend money buying in queens from elsewhere. They then wonder why the special qualities for which they bought the queen don't last into a subsequent generation and so they have to go back to the vendor again next year. The beekeeper would do better to select from his own best and to supply his neighbors cheaply or for free with his queens to discourage them from buying in unsuitable genes. After a few years of selection (culling the poor ones to ensure they don't produce drones) you should have a local strain well-suited to the area. Of course, this depends on knowing whom your neighbors (a drone flight x2) are and this is another reason for joining your local association!

Beekeeping with the Seasons

General Comments

One should start planning as early as possible. For instance, in the late fall or early winter you should be planning what type of honey and other bee products you will be producing; procuring and assembling the equipment, and generally getting ready for the spring flow. If you are going to purchase packages , you should be contacting the package and other bee producers early since bees are in short supply. Better still, you should be planning to produce nuclei or packages for your neighbours, complete with queens that you would wish to form part of the local gene pool. You should also be ordering bee equipment such as veils, smokers , hive tools, etc. in addition to ordering a quantity of jars in the late fall, early winter since they are cheaper. Normally, my goal is to have the new hives assembled and painted ready to go one month prior to bee installation and the supers ready to go one month prior to the honey flow as the flow may come early and you will not be ready.

Starting early particularly applies to the bees you want to work with next year. Your record cards (or your memory of bad stinging events) will tell you which queens you want to be contributing to the gene pool next year and, more importantly, which ones you would rather were eliminated. By the time you decide, probably this year's queen rearing and mating will be over. At the end of the honey gathering season unite "bad" colonies to "good" ones, making sure that the "good" queen survives. The united colony will enter, endure and leave winter much stronger and, come the spring, will be in good condition for early queen rearing and division. Another advantage is that, if you are in the habit of feeding your bees, you will have fewer into which you feel the need to pour 'chemical' sugar. The message bears repeating that Chris hasn't routinely fed sugar to his bees for years and regards this as part of a Darwinian selection process to eliminate those strains that are unfit for their local environment through inefficient overwintering. Saves a lot of unnecessary work and expense too!

Early Autumn (Fall in the US) is a good time to add propolis screens or mesh to collect a supply of this valuable product. It is now appearing in all sorts of products: tinctures, sweets and toothpaste are but three examples.

You also need to plan bee yards and make the necessary agreements early to locate your colonies on someone else's land.

TYPICAL SEASONAL MANAGEMENT

Mid-State (Columbia) South Carolina, USA Area

Winter (December thru February)

1.0.1. December
- Start thinking about how many packages and queens you want to expand to and who to order from
- Continue repairing / painting old equipment and assembling / painting new equipment
- Make arrangement for any new beeyards
- Contact property owners and make agreements

Get special labels printed for honey and hive products for Christmas gifts

If your hives are brood free (and if it is legal where you are) treat the bees against varroa with a dose of oxalic acid (oxalic acid not currently legal in the US).

1.0.2. January

- On warm days in second half of January go thru hives and determine whether disease free, food – pollen and honey - are sufficient and they are dry:
- if they need feeding assess Varroa mite levels and treat if necessary with an appropriate IPM measure.
- treat for AFB or EFB if present – Terramycin, etc. , Better still, eliminate AFB by destruction by fire of the entire contents of the hive and scorching the interior of the hive with a blow torch.
- check for laying queen and brood present.
- put initial order in for packages and queens
- reverse brood chambers if needed to get the queen laying in the bottom brood chamber and as a swarm prevention technique.

1.0.3. February

- continue feeding if needed
- mid to latter part of February, start swarm prevention measures, especially on hives feeding or
- with a lot of stores.. Decide what you will do with surplus syrup-derived stores to prevent them contaminating the purity of your honey.
- put final order in for packages and queens or turn the swarming instinct to advantage by producing packages and queens for your own use and that of your neighbours' queens - plan which hives you want to split, requeen, combine.
- remove entrance cleats
- reverse brood chambers if needed; again consider whether to make use of the frames and bees in one of the chambers to make increase for use or for sale.

Spring (March thru May)

2.0.1. March
- Check for laying queen
- check for diseases
- swarm prevention Make a shook swarm to put the bees onto fresh foundation or starter strips. This is best done at dandelion time. In bad weather you may need to feed. The bees will romp ahead on clean comb. If varroa is present then sacrifice the first brood to be sealed: this will get rid of most of the mites.
- Put on Queen Excluder (if you use them) AND supers mid to late March.
- First of March, pull all strips out, patties, etc.
- Put on supers mid to late March.
- A honey barrier will keep the queen from moving upward; but is not as effective as a queen excluder. It seems to be the American practice to extract honey from combs that have been bred in, whereas in the UK and Ireland this is usually frowned upon, possibly because the beekeepers know what lies beneath the old cocoons and is released by the uncapping knife. In the American market in particular, where light honey is esteemed, the queen should be kept out of the supers as the honey when subsequently extracted will be darker.
- Place section supers above brood nest, especially if you want to harvest some delicious comb with bee bread
- 8-9 frames in extracting supers
- rotate center frames to outside.
- top supering, bottom supering
- removing bees from supers
- brushing, chemical fume board, bee escape boards.
- Make sure the hives have enough honey SIGNS OF HONEY FLOW whitening along edges of combs

- bees flying vigorously in and out of hive
- see fresh nectar in cells

2.0.2. April
- go in and check for swarming ; perform swarm prevention.
- make sure have enough supers, better to over super than under super.
- can make splits if desired
- Undertake queen rearing from a selection of your best queens .

2.0.3. May
- check for swarming ; perform swarm prevention.
- make sure have enough supers, better to over super than under super.
- begin plans for extracting: number of supers to extract and which beeyards to extract in what order
- can make splits if desired
- make sure you leave enough honey on for breaks in available forage.

Summer (June thru August)

3.0.1. June
- pull supers and extract.
- 3/4 – 7/8 of super is capped depending on when in season. In spring and early summer this rule should be adhered to since the honey has a higher water content. In fall, uncapped honey is probably ripe and can be harvested.. A simple test it to shake the comb upside down. If liquid drops out it isn't ripe honey. Or buy a refractometer to measure the moisture content; their price is coming down.
- can make splits if desired.

3.0.2. July
- check for stores, diseases .
- swarm prevention
- assess the new queens you have reared for brood pattern to ensure they were well-mated.
- or order requeening queens for fall requeening

3.0.3. August
- treat for diseases if present
- start thinking about what hives may need to combine for winter or order another queen and requeen weak hives.
- requeen

4.0 Fall (September thru November)

How your hives are this fall will determine how successful you are with your honey next spring.

4.0.1. September
- combine hives if needed: rules for combining: weak/weak; strong/weak, strong/strong.. Combine only healthy colonies as one unhealthy colony combined with a bigger one results in a bigger sick colony.

4.0.2.October
- make sure they have plenty of food; feed sugar syrup if needed.
- begin ordering new equipment you will need next spring
- repair and paint old equipment
- remove queen excluders

4.0.3. November
- assemble new equipment and paint
- begin making plans for spring
- repair and paint old equipment
- Reduce main entrance

- Winter cluster movement thru the winter.
- Water vapor that must be allowed to escape
- Survival depends on:
    » young, vigorous queens
    » large population of bees
    » adequate supply of honey
    » disease - mite free
    » top ventilation unless on mesh floors

Some places, for example the northern USA and Canada, may have climate and predators less benign than in Dave's area. For instance there might be harder winters and there might be bears! Here's how they cope with both in Slovenia. The pictures are courtesy of Franc Sivic, the beekeeper who won a Gold Medal at the Melbourne Apimondia for his photographs of bees on flowers.

Figure 73 Slovenia Bee House
The beehouse provides much better warmth and shelter from the elements than does the standard Langstroth hive, even when wrapped. Notice the wire mesh panels to keep bears at bay. Here's a close up:

Figure 74 Slovenia Bee House Close Up

Moving Bee Colonies

Colonies in the summer should be closed early in the morning or late at night when all the field bees are in the hive. In the winter, the low temperature should be closely monitored to determine if you will get chilled brood due to moving (i.e. the bees break cluster and do not cover the new brood after the shortest day in December.) If not on a mesh floor, screens can be cut to the width of the front entrance and folded to press fit into the front opening to keep the bees in the hive. Moving screens and "front porches" can also be utilized. A strap should go around the colony or duct tape may be used. Duct tape should be utilized to close any openings and other holes in the hive. One person can lift a single deep colony but two deeps or a deep and shallow or medium super should be lifted by two people. Alternatively, a hand truck can be used to lift the deep + super colony by one person. In cool weather the bees may be using only the bottom box and so the hive can temporarily be reduced to that for easy lifting.

Alternatively, the hives can be all loaded into the truck and a net put over the entire load.

Note how the bees will propolise a moving screen if left on the hive. However, in very warm climate a moving screen may provide additional ventilation necessary.

A trailer with a drop down tail gate contacting the ground can be utilized with the hand truck carrying the hive moved up the tail gate onto the trailer. Minimal lifting is required with this technique. Often, for a hobbyist, renting a trailer may be more cost effective.

Figure 75 Moving Screen
Dadant Moving Screen

Figure 76 Propolised Moving Screen

If the bees are clustered on the outside bottom board (stoop) of the hive, smoke them a couple puffs of smoke, give them a few minutes, and they will go into the hive.

Use duct tape or beeswax instead of nails whenever possible. In conditions such as the moist South East United States nail holes from removed nails will rot, shortening the woodenware life.

When loading colonies, keep the engine running and the vibrations will help keep the bees calm.

A pickup truck can improve the efficiency of the beekeeping operation. For small operations, say less than 200 colonies, a ½ ton class truck with a flat bed does well. A standard ½ ton truck has the problem that one cannot reach over the side of the truck to get equipment out of the bed. A small truck does allow a beekeeper with a small operation to reach into the bed of the truck. An extended cab with the bench back seat with the small fold out door is nice to place bee equipment such as veils, smokers , straps, extra fuel for the smoker such as pine needles in.

A Langstroth colony weighs between 125 and 175 pounds depending on the configuration. Say the average is 150 pounds. Nine colonies will fit one colony tall, in the back of a standard pickup. This is 9 x 150 = 1350 pounds. So, the pick-up needs to be able to handle 1350 to 1500 pounds. Refer to the on-line dealer specification sheet for more information and the latest information.

Trailers need to also be specified with respect to the towing capacity of the truck. The reader should consult the on-line dealer specification sheet; determine how many colonies he wants to move, to determine the proper truck / trailer combination. However, for many small operations, a small truck will suffice. It get's better gas mileage also. Chris was getting about 57 miles per gallon of diesel from his Vauxhall Astra Estate and this has gone up to over 60 mpg since he was fined for speeding. He had a trailer made to fit 3 hives across by 4 long, but it is rarely used as it is easier to put hives in the back of the car.

Extracting Honey

Honey may be taken off the hive by moving the bees out of the super by chemicals (e.g. Bee Go or Bee Quick in the US), brushing or shaking the bees off the frames, escape boards and the old fashioned Porter bee escape. Escape boards work very well but require two trips, one to put the board one and another to pull the honey. Chemicals require only one trip and are quicker but are not as reliable.

Figure 77 Bee Escape Board

Brushy Mountain Bee Farm bee escape board. Upside down in the picture. Bees go down the hole and out thru the openings in the triangle's apex. They cannot figure out how to go back thru the triangle, trying instead to get through the mesh below the hole.

Figure 78 Fume Board
Dadant Fume Board

If you don't want to go to the expense of a fume board, try stapling, with the office stapler, material, such as butter muslin, to black plastic taken from a refuse sack (unused!).

Figure 79 hand Powered Extractor
Brushy Mountain Bee Farm manual extractor

Honey, once taken from the hive, should be extracted as quickly as possible to prevent wax moths and small hive beetles (not in Europe yet except in Government laboratories as far as is known) from multiplying in the unextracted honey and comb and to reduce the amount of honey crystallizing in the comb. After extracting, the wet supers still have a thin coating of honey on the combs and should be placed back on the same colonies they came off. Placing supers back on the same colonies is a disease reduction technique as it avoids spreading any, so far unnoticed, disease that might be present. Also, if one wants to avoid using Paradichlorobenzine (PDB) to keep wax moths out of comb, brood should be kept out of extracting supers. PDB leaves residues which are illegal in some places and undesirable in all.

The extracted supers with white comb can be stacked criss cross over winter to let air and light into the supers. A queen excluder top and bottom will keep mice out. Wax moths are after the dark comb which occurs from brood rearing. In places where there isn't much of a winter, wax moth damage can be minimized by placing whole supers, one at a time, into a deep freeze for a day or two, then removing and placing into a plastic bag which is then sealed to prevent wax moth entry and labeled. Some beekeepers have a deep freeze dedicated for this purpose.

When extracting large numbers of supers, a double buffer approach is most efficient. Double buffering is where two smaller extractors should be used. When filling up one extractor, the other extractor can be extracting the honey. On the other hand, is it better to have one or to have two cumbersome items that are expensive space-wasters for 364 days a year? This is yet another reason to join your local BKA which may have one to borrow or to hire.

Radial, center drain extractors are recommended. The Radial is the quickest and most efficient since the frames do not have to be reversed, over the tangential extractor where the frames have to be reversed. The center drain extractor is better over the side drain extractor to get that last bit of honey out.

Bee Space

When purchasing equipment, a good rule of thumb is to purchase all your equipment from the same vendor. In the case of Langstroth equipment, you are concerned about the bee space between the bottom of the "top" super to the top of the "bottom" super. If this is violated due to mixed manufacturer equipment, the bees will either build burr comb in the space if it exceeds the bee space (1/4" to 3/8") or propolise the space closed if it is less than the bee space. The US bee magazines such as American Bee Journal or Bee Culture periodically carry charts with various manufacturers' equipment. From these charts/dimensions you can tell what manufacture's equipment will work with what other manufacturer's equipment.

In addition to the bee space between supers, you are concerned about the outside dimensions of the equipment and the supers fitting on top of each other. This is a concern but not as big a concern as the bee space.

Colony Numbering

Often a beekeeper needs to number his colonies to keep track of colony status. Numbers should not normally include things like queen pedigree, strength, etc. This information should be written in a ledger not on the colony. It is recommended a straight number scheme be employed. The simpler the better with keeping in mind if you write the number on the bottom brood chamber (Langstroth) if you switch out the bottom brood chamber you so need to make a note about the change in numbers in your ledger. However, the system Chris employs of the hive bearing the number of the queen, allows both colony and pedigree tracking on the record cards.

If you are curious about the numbering scheme used on the equipment in some of the book's pictures, the first two digits are the last two digits of the year the equipment was put in service, M stands for medium super, s stands for shallow super, and B stands for brood chamber, the last two digits to the right stand for what number the equipment was placed in service that year. I use this for accounting purposes and to identify diseased equipment prior to thorough cleaning or destruction by fire.

Eight Frame and 10 Frame Equipment

For 8 frame and 10 frame equipment, weight and being able to get your arms around the super are important items. A deep 10 frame super full of honey (this is the same box as a deep brood chamber but just placed above the brood nest to collect honey) weighs up to 60 – 80 pounds (1) (19 3/16 x 16 1/4 x 9 7/16 inches), with a 10 frame medium 6 5/8 inches super weighing as much as 45 pounds (19 3/16 x 16 1/4 x 6 5/8 inches), a shallow 10 frame super will weigh up to 35 pounds (19 3/16 x 16 1/4 x 5 11/16 inches), and 8 frame medium super is approximately 36 pounds (19 3/16 x 13 3/4 x 6 5/8 inches). Of note is that 8 frame equipment is 13 3/4" deep compared to 16 1/4" deep of 10 frame equipment. This makes a big difference in being able to handle the super (force = super weight x super center point distance). In addition, the 8 frame medium super weighs approximately nine pounds less than a 10 frame version.

On a recent trip to America Chris noticed that most people seem to have less distance between navel and fingertips than generally is the case in Europe and this may explain the preference for the narrower Langstroth hive over the British National which is 18 1/8" square.

Figure 80 Eight Frame Brood Chamber
Brushy Mountain Bee Farm 8 Frame Brood Chambers.

Figure 81 Ten Frame Brood Chamber
Brushy Mountain Bee Farm 10 Frame Brood Chamber

Often bees find it difficult to draw out the outer most 1 to 2 frames on each side of a 10 frame super. Eight frame equipment minimizes this problem and supports the bees' natural tendency to build up more than to build out. However, if commercial production is desired, the 10 frame equipment is cheaper/cell (1). It should be noted that the brood area should still be allowed to mostly fill prior to placing extra supers on top of the brood chamber.

The 'National' hive used by 80% of the beekeepers in the UK and Ireland takes 11 frames in the brood box and there is room for a 'dummy board' also if Hoffman spaced frames are used.One great advantage of the National hive is that the boxes are square and so, if desired, they can be stacked crosswise. This reduces the tendency of bees (especially if beespace isn't quite right) to build wax bridges from the top of one comb to the bottom of the one above.

Nucleus Colonies nucs (or nuclei)

Managing your colonies with nucs is excellent. As the great E.B.Wedmore emphasised in his Manual of Beekeeping in 1932 "Almost every emergency of management can be met forthwith by putting something into or taking something out of a nucleus, while nuclei themselves seldom present emergencies." Nucs are four or five frame boxes that you put frames in that you may pull out of ten frame colonies for swarm reduction. Nucs are easily combined with larger colonies that have a poor queen, may have lost a queen or need requeening for other reasons. Requeening with nucs usually has a higher success rate than requeening with the Benton Cage that you get from queen producers. Nucs can be overwintered easily in the south and are usually consolidated into groups to get thru the winter in the North.

Figure 82 Nucleus Box
Dadant Nuc

Nucs seem to work better for small quantities of bees than the ten frames Langstroth. The frames from a Nuc are easily transferred to an eight or ten frame brood chamber. Nuc supers are available as is the Queen Castle, that I helped design, from Brushy Mountain Bee Farm in the US. The Queen Castle is a ten frame brood chamber that can be divided into two, three, or four nuclei.

When purchasing nucs you need to be careful about diseases and also not getting bad combs. Most bee suppliers are probably alright most of the time with respect to diseases but individuals may not have the experience that the bee suppliers do. Bad combs may be comb where the cells are not fully developed or malformed due to heating, etc. Bad combs will have to be rotated out of the colony or nuc. Bad combs also include comb where the bees have rebuilt the comb in an empty space in the comb. This empty space may have been due to wax moths or small hive beetles. Usually the bees will build drone comb in this empty space since most colonies do not have 14% to 17% drone comb that occurs naturally.. However, Chris has found that where he has deliberately removed the bred-in area of old combs the resulting naturally-drawn comb often has a remarkable good brood pattern with very few 'misses' in the sealed brood. This is especially true for Top Bar Hives and hives with foundationless frames (frames drawn out naturally without foundation as a guide).

Business Aspects of Beekeeping

There is a cost , volume, and distance relationship in the bee business. For those beekeepers with colonies in their backyard (garden), they can afford to make multiple treatments of their colonies since they do not incur travel cost. This is important in making decisions as to which medications to treat with.

If the beekeeper has outyards (apiaries), the greater the numbers of colonies in the outyard, mean the least amount of travel cost per colony. In addition, it costs more to travel greater distances, which means more colonies would need to be involved to cover the travel cost. This is why tractor - trailer trucks (lorries) and a large number of colonies are moved greater distances.

On the other hand, the greater the number of colonies in a yard, the greater the competition for available resources. Some have found that the total honey crop from 10 hives on a site to be no more than with 5 hives on the same site in other years. Bailey (United Kingdom) tells us that if there are too many hives for the available forage, the foragers will spend more time in the hive leading to increased transmission of tracheal mites (Acarine in the UK) and Chronic Paralysis Virus. It may also lead to less efficient distribution of queen substance, leading to swarming or to super-sedure. It is therefore a very local issue that is different depending on the flora and geography.

Placing beeyards

Beeyards should be clustered within 4-5 miles of each other to conserve fuel cost. This 4-5 mile distance is twice the maximum distance that bee will generally fly, so avoiding overlapping forage areas and unnecessary competition. It makes sense to work out a route, for example a circuit that you can manage in an afternoon, with an apiary spaced every few miles along the road.

Determine the type and quantity of flora in the area. Each yard should have the maximum number of colonies that the amount of flora in the area will support throughout the year and also the number of colonies that a beekeeper can support /manage. Consider the cost, volume, distance relationship. If you have too many hives on a site you will have extra management costs through having to feed or to move colonies, as well as increased likelihood of pest and disease problems and greater losses if the law where you are insists on destruction of colonies with AFB.

A beeyard should be located such that your truck may be driven up to the back of the colonies or in close proximity. Honey supers (if you're lucky) get heavy and trying to move them can bring on 'beekeeper's back'; in addition, having the truck close to a colony that is being moved sure makes life easier.

Equipment

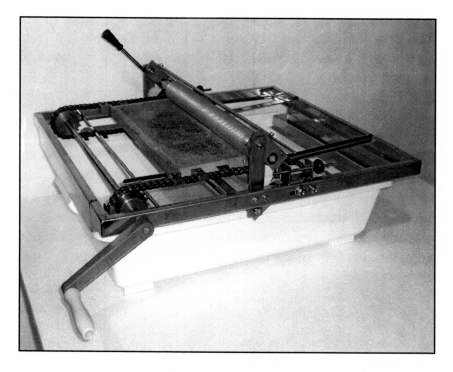

Figure 85 Hand Uncapper
Brushy Mountain Bee Farm hand uncapper so keep your hands well clear.

When should an uncapper be purchased? If you save one minute per frame and the labor rate is $10/hour, and the cost of the uncapper is $2,000:

1 Frame/minute x 60 minutes/hour = 60 Frames/hour or approximately 6 supers/hour

Let's determine the number of colonies that is required to justify the uncapper: The .8 means that 80% of the colonies are producing.

X colonies * 2 supers/colony * 10 frames/super * .8 * 1 hour/60 frames * $10/hour = $2,000
X = 750 colonies.

If we drop the uncapper cost down to $1200, we have:
X colonies * 2 supers/colony * 10 frames/super * .8 * 1 hour/60 frames * $10/hour = $1200
X = 450 Colonies.

One should note that increasing the labor rate or increasing the time saving per frame will reduce the number of colonies required for breakeven. However Chris recently timed himself using a carving knife with a serrated edge for uncapping and it took about 30 seconds a frame. The carving knife cost a couple of pounds and occupies very little space when not in use. The message is 'keep it simple.'

Hive Stands

The reasons for using hive stands are:

- To get the colony off the ground so the wood will not rot and decay quickly

- Keep bugs out

- To keep small animals out of the colony

To assist in making the colony an optimum height to work, the colony should not be more than 6-12 inches off the ground.
 If stacking boxes, e.g. Langstroth or National hive is used, allow for the height when supered afterwards. A top bar hive or similar can be placed on a trestle stand so that the top of the hive is at waist level thus avoiding beekeeper's back.

Figure 87 Cement Block Hive Stand
Note cement blocks for a hive stand.

This Portugese beekeeper uses wooden rails supported on blocks on which to stand his hives.

Figure 88 Portuguese Beehive on Wooden Stand

Notice that the plastic floors have indentations for the purpose. They also have a couple of holes in the front to enable an alighting board to be attached if desired. The disadvantage of rails, of course, is that the hives have to be kept in straight lines which encourages drifting. Drifting means that any disease is quickly spread and that guards are constantly on the alert for potential robbers which may make them appear more defensive (or aggressive) than would normally be the case. Also there may be a tendency for bees to congregate in a few hives and not evenly. This can be turned to advantage by using the hives as a set of stairs!

Figure 89 Portuguese Hives

It saves the 'vertically challenged' from taking a step ladder to the apiary.

The hives shown above are pretty new. In the hot, humid and rainy climate of North Portugal, Mr. Jorge, the beekeeper pictured above, has hit on the idea of extending the life of his hives by stapling to the outsides opened -out foil-lined drink cartons! He reckons that it extends the life of a brood box from about 5 years to up to 15. You can see the problem from the picture below. Mr. Jorge is what the Americans call a 'sideliner' with about 200 hives.

Figure 90 Portuguese Hives
Note stand and telescoping top

The hives are very cheap compared to UK prices, and are made from imported Russian pine. The hives stacked in the bee appliance factory pictured below show plenty of knots in the wood.

Figure 91 Portuguese Equipment Factory

Figure 92 Power Zone

This hive stand is three cement blocks with two treated landscape timbers on top of them. Notice how I can stack extra equipment on the hive stand. I typically do not place more than two colonies on a landscape timber stand at a time and this is in non-African Bee area. In this photo note that the second brood chamber is in Cliff's lower power zone and the top super is starting to be too tall for Cliff's upper power zone. At this point the beekeeper should consider removing the top supers above the power zone and possibly place them on other colonies if he did not want to harvest the honey yet.

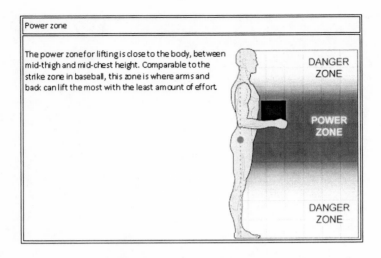

Power zone

The power zone for lifting is close to the body, between mid-thigh and mid-chest height. Comparable to the strike zone in baseball, this zone is where arms and back can lift the most with the least amount of effort.

DANGER ZONE

POWER ZONE

DANGER ZONE

Figure 93 Human Power Zone

You want the hive stand to be of a height that supports lifting in the power zone. Let's look at some dimensions:

If the hive stand is	12 inches off the ground,
and a deep brood chamber is about	10 inches
another deep brood chamber is about	10 inches
A medium super is about	7 inches

The total hive stack for this configuration is 39 inches or 1 yard (3 feet) and 3 inches. The second deep brood chamber is in the power zone of most people. One would have to measure from the floor to mid-thigh and from the floor to mid chest to determine their power zone distances. Then compare these distances to the hive configuration you will be using, for a Langstroth deep 9 7/16", for a medium 6 5/8", and for a shallow 5 11/16" or round the dimensions up like I did. Having the hive stack match your power zone is important not only for safety, but also for work efficiency and speed.

If you already have a back problem or don't relish the thought of lifting heavy boxes in any case, then think laterally and go either for a top bar hive or one of Robin Dartington's long deep hives (http://www.dartingtonhive.co.uk/). The stand can then set the top bars at about navel height for easy handling without stooping. Even if the Dartington hive is supered the full supers weigh only about 16lb each. A full frame of honey from the brood box weighs about 8lb.

In warmer climates, two landscape timbers on two to three cement blocks work well and are easy to set up and carry. Extra equipment can be stored on the extra space. Two cement blocks can also be used to set a colony on. Two cement blocks are the cheapest but there has been concern about moisture seeping up the block. This seems to not have been an issue in use in the South Eastern US (South Carolina). If you are concerned about moisture, place some roofing felt paper (tar paper) or old roofing asphalt shingle on top of the block before placing the hive on the block.

Also, if more than one hive is on a stand, there is concern about one hive irritating the other hive when working the colonies. This is especially true for Africanized bees. It is generally recommended to have one hive per hive stand; two at most. Smoke both at the same time.

Colony Covers

Figure 94 Mann Lake Migratory Cover

Figure 95 Telescoping Cover
Brushy Mountain Bee Farm telescoping top. Dadant is almost identical

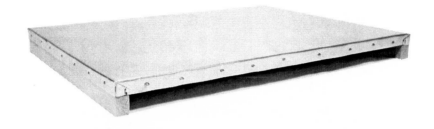

Figure 96 Migratory Cover
Brushy Mountain Bee Farm Migratory Top, covered

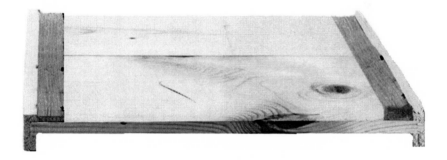

Figure 97 Florida Migratory Cover

In the US, the aluminium coated telescoping cover is what most hobbyist use. This cover last a long time and is very durable. However, when moving colonies, especially in the winter, this cover does not "stack" well in the back of a pickup truck. The left and right cover sides waste valuable space. In the summer, the telescoping cover should be removed and a moving screen placed on top of the colony.

Migratory covers are the cover of choice for a lot of commercial beekeepers. They are cheaper than telescoping covers but do not last as long if they are not coated with aluminum sheeting. Migratory covers are flush with the left and right sides of the colony and stack well in a pickup truck. One issue with migratory covers is the super frames sticking to the cover bottom if an inner cover (crown board) is not used. Beekeepers use the migratory cover with and without the inner cover depending on preference.

Figure 98 Standard Ten Frame Hive

Note this configuration. It has a Dadant Florida migratory cover, then a Brushy Mountain Bee Farm inner cover , then a Dadant brood chamber, then a queen excluder, the Dadant bottom brood chamber, and the Brushy Mountain Bee Farm bottom board. I do not normally run a queen excluder but let the queen expand her brood nest normally in the spring. Often, I will run a Deep Brood Chamber, a medium super food and brood chamber, then my honey supers.

Index

Glossary of Terms

Truck: a Lorry

Lift : a hand truck

Barrow : a wheel barrow

Sledge : toboggan

Kit: a bee hive

Langstroth: a Reverend gentleman of that name devised and made popular the forerunner of the moveable frame hive which essentially consists of a set of stacking boxes. Thus he has inadvertently been responsible for innumerable cases of 'beekeeper's back' over the last 150 years or so.

Dadant: A French emigrant to the US and a contemporary of Langstroth. He designed the hive that bears his name and founded the company that exists and thrives today.

Fall: Autumn

Yard: apiary

Outyard: out apiary

Screened bottom board: open mesh floor

Landscape timbers: treated or Tanalised wood.

Inner cover: crown board.

Sideliner: beekeeper on a scale that makes a significant proportion to the family income but not all of it.

Michaelmas: the Feast of St Michael and All Angels – 29th September. It is around this time that the Michaelmas daisies bloom.

Bailey: Dr Leslie Bailey, formerly of Rothampstead Experimental Station (UK) who did much work on bee viruses.

Eke: extension, extra rim used as a spacer.

Corn: maize

Pail: bucket

Slanticular: aslant, not perpendicular to the ground.

Nice: good but not outstanding.

Honey flow: a fortunate combination of weather, bees and flowers enabling much nectar to be gathered in a short period. Apparently they come by the calendar in the US but such occasions are rare and unpredictable in Europe!

Index of sources

(1) "Which Supers to Choose," David MacFawn, American Bee Journal, February, 1995 pp101-102 Volume 135 No 2.

(2) "Effect of Population Size on Brood Production, Worker Survival and Honey Gain in Colonies of Honeybees," John R. Harbo, Journal of Apicultural Research 25(1): 22-29 (1986).

(3) "Effect of Comb Size on Population Growth of Honey Bee (Hymenoptera: Apidae) Colonies," John R. Harbo, J. Econ. Entomology: 1606-1610 (1988).

(4) "Worker-Bee Crowding Affects Brood Production, Honey

Production, and Longevity of Honey Bees (Hymenoptera: Apidae)," John R. Harbo, J. Econ Entomology 86(6) 1672-1678 (1993).

(5) "Bee Research Digest, Edward E. Southwick, American Bee Journal, pp349-350, May 1995.

(6) Email correspondence between John R. Harbo and David MacFawn," Bee Density," August, 2006

(7) "The Comb Honey Book," Richard Taylor, Linden Books, Interlaken, N.Y. ISBN0-9603288-0-7

(8) "Honey in the Comb," Eugene E. Killion, Dadant & Sons, Hamilton, Illinois ISBN 0-915698-08-0

(9) "The Hive and the Honey Bee," Dadant, ISBN 0-915698-09-9 p738, 178

(10) ABC and XYZ of Bee Culture," A.I. Root, Medina, Ohio ISBN 0-936028-01-7

(11) Plastic Comb Foundation Can Hinder Comb Building and Honey Production," Thomas D. Seeley, pp 955-957, American Bee Journal, November 2006 Volume 146 No. 11

(12) U.S. Department of Labor
Occupational Safety & Health Administration
http://www.osha.gov/SLTC/etools/electricalcontractors/ma-terials/heavy.html#Weight%20of%20Objects
www.osha.gov

(13) G.M. Doolittle, " Scientific Queen-Rearing," 2008 Reprint

Wicwas Press, 1889, 1899 George W York, American Bee Journal, 118 Michigan Street. ISBN 978-1-878075-24-6

(14) Giles Fert, " Breeding queens," Translated in English by N. A. Riley and P. M. Greenhead O.P.I.D.A.ISBN 2 905851 11-2

(15) Harry H Laidlaw and Robert E. Page, "Queen Rearing and Bee Breeding," Wicwas Press, Cheshire, Connecticut, USA original 1907 ISBN 1-878075-08-X

Breinigsville, PA USA
23 March 2011
258254BV00003B/167/P